CALGARY

Metropolitan Structure and Influence

CALGARY

Metropolitan Structure and Influence

edited by

BRENTON M. BARR

Western Geographical Series, Volume 11

Department of Geography
University of Victoria
Victoria, British Columbia
Canada

1975 University of Victoria

Western Geographical Series, Volume 11

EDITORIAL ADDRESS
Harold D. Foster, Ph.D.
Department of Geography
University of Victoria
Victoria, British Columbia
Canada

Publication of the Western Geographical Series has been generously supported by The Leon and Thea Koerner Foundation, the Social Science Research Council of Canada, the National Center for Atmospheric Research, the International Geographical Union Congress, the University of Victoria and the National Research Council of Canada.

ISSN 0315–2022
ISBN 0-919838-01-4

ACKNOWLEDGMENTS

This volume is the second in the Western Geographical Series to be devoted to a Canadian urban area. As such it has an orientation relevant to one of the three areas of specialization developed by the Geography Department of the University of Victoria.

Many people have combined their efforts to facilitate its successful production. Dr. Brenton M. Barr has devoted both time and effort to its editing. The layout and design of the cover and contents, together with much of the photographic work was undertaken by Mr. Ian Norie assisted by Mr. Ole Heggen and Mr. Ken Quan. Mrs. Alison Griffith also provided valuable assistance. The manuscript was typed by Mrs. Laurel Carr and Miss Louise A. Evans. Their kind cooperation is gratefully acknowledged.

Harold D. Foster

University of Victoria
Victoria, B.C.
September, 1975

PREFACE

Throughout the twentieth century, Calgary has been synonymous with the exploration for, and development of petroleum in Canada. Although the city serves provincial and regional needs for many commodities and services, most of its growth in this century in some way has been associated with different periods of growth/in the western petroleum industry. Now, in the 1970's, as exploration and investment in the Canadian conventional petroleum industry shifts to arctic and eastern regions, the future role of Calgary appears to hang in the balance; at this point in time, therefore, an examination of Calgary's established metropolitan role and structure is appropriate so that the implications of changes in the conventional petroleum industry and in the Northern Alberta oil sands can better be appreciated as they begin to exert spatial influences on the organization of urban-economic processes and patterns in Canada.

As the title implies, two major themes are contained in this volume: the first four chapters examine aspects of Calgary's spatial influences as a western metropolitan complex; the last three chapters focus on Calgary's internal socio-economic structure. Together the two parts of the volume offer exciting perspectives on the role of Calgary as a metropolitan complex and the internal spatial organization which has evolved as the city's economic prowress has grown. All seven chapters in this volume have been prepared from research findings obtained by the authors as part of their major investigations into specific facets of Calgary's spatial structure.

Calgary's role in the Alberta manufacturing economy is investigated in Chapter 1 by Brenton M. Barr who has examined inter-industry manufacturing linkages of Calgary firms within the provincial economy for evidence of the existence of a provincial and metropolitan industrial complex. The same author and his colleague, Alice McEwen, examine in Chapter 2 selected aspects of intra-industry manufacturing linkages between Calgary firms and firms located in southern Alberta which are engaged in the mobile-home industry; this study investigates the extent to which functional subordination exists between metropolitan manufacturers in Calgary and those located in smaller urban places in the Calgary region. Calgary's role and influence as a centre of tertiary and quaternary activity in the Canadian and international petroleum industry is documented by George Zieber in Chapter 3. While Calgary's role in petroleum administration is generally known, the city's important role in petroleum operations is usually underestimated. The final chapter in Part I comprises an examination by Wayne Davies and Tim Gyuse of post-World War II changes in the central place system in the Calgary region. The first four chapters of this volume offer, therefore, new and important insights into Calgary's spatial influence on secondary, tertiary, and quaternary economic activity.

Part II contains three chapters which analyze aspects of Calgary's internal socio-economic structure. Peter Smith and Don Harasym examine, in Chapter 5, how planning policies have affected the form, function and spatial arrangement of retail services in the new residential districts, which have evolved during the past thirty years, as the city's population has more than quadrupled. Denis Johnson's study, Chapter 6, nicely complements the previous chapter by comparing aspects of food-store and dwelling linkages in two older residential areas of the city. Wayne Davies, in Chapter 7, offers an appropriate summary to the volume in his measurement and description of the social and demographic character of Calgary communities on the basis of 1971 census data.

The seven chapters of this volume represent the major themes and avenues of research associated with economic and socio-economic geographical analyses of Calgary by geographers in Alberta. While this volume focuses on specific issues and topics, the editor hopes that analyses of Calgary, in the future, by economic and urban geographers will probe many of the unresolved issues presented here and will greatly broaden our understanding of Calgary's role as a metropolitan economic structure.

Brenton M. Barr

University of Calgary
Calgary, Alberta
May, 1975

TABLE OF CONTENTS

LIST OF TABLES

LIST OF FIGURES

LIST OF PLATES

PLATE 1
The Calgary Stampede.

Calgary Exhibition and Stampede Photo

PLATE 2
Fort Calgary, 1875.

PLATE 3
Blacksmith's shop, Fort Calgary, circa 1875.

Glenbow Foundation Archive Photo

SECTION I

SPATIAL INFLUENCES AS A WESTERN METROPOLITAN COMPLEX

"Calgary as a manufacturing system,
however, is not primarily dependent
on sales within Alberta".

"... nearly eight percent of Alberta's
manufacturers are members of the
mobile home industry".

"In spite of the fact that oil activity
is spreading out and becoming
national in scope, Calgary still is
the base for the directing of activities".

"In general, the changes found
over the period show the familiar
story of the increasing centralisation
of commerical functions in the
larger rural service places... "

PLATE 4
Calgary's City Hall, constructed of sandstone, 1907.

City of Calgary Photo

CHAPTER 1

THE IMPORTANCE OF REGIONAL INTER-INDUSTRY LINKAGES TO CALGARY'S MANUFACTURING FIRMS[1]

Brenton M. Barr

University of Calgary

INTRODUCTION

The Problem

The industrial prowess or magnitude of production of urban manufacturing systems is usually inferred from criteria such as numbers employed, the number of industrial establishments, value added through manufacturing, and selling value of factory shipments. Data to measure and assess these criteria are reasonably accessible for most urban areas in Canada and are collected and published by federal and provincial governments. The functional importance of Canadian urban manufacturing places, however, cannot be measured directly from published data since these data are neither collected universally by government nor published in any consistent manner by other agencies. The general economic relation of any urban manufacturing economy to other urban or regional manufacturing complexes must at best be inferred from piece-meal data or through statistical estimations of expected interaction, (see for example, Britton, Gilmour, Karaska).[2] Except in those cases where dependency is clearly demonstrated by such phenomena as shipments of raw materials from unifunctional sites to specific manufacturing complexes,[3] the general problem of industrial interconnectivity is very difficult to assess with any degree of reliability. Solution of this problem is, however, important if a realistic assessment of the functional relationship between urban places and their hinterlands, or urban places and industrial enterprises in other urban spaces is to be made.[4] Assessment of the functional basis of any urban place is mandatory before realistic and productive policies and programs can be adopted and implemented to accomplish such objectives as economic stability or employment growth.

Objectives

The general purpose of this paper is to describe and assess Calgary's industrial role within the Alberta provincial economy through examination of patterns of inter-industry linkages displayed by manufacturing firms operating in Calgary. These manufacturing linkages are specifically examined to provide evidence of Calgary's role:

1. As a consumer of provincially-manufactured goods.
2. As a producer of provincially-consumed products.
3. As a member of a provincial manufacturing complex.

Since Calgary is popularly understood to be a regional distribution and gathering centre and the administrative and technical capital of Canada's petroleum industry, analysis of inter-industry manufacturing linkages here is intended to further the general understanding and appreciation of Calgary's role as a consumer and producer of manufactured commodities. This paper is directed toward those interested in the general academic study of urban manufacturing complexes, those concerned with the impact of resource-extraction on the dominant urban places in a sparsely-populated region, and those charged with creation and/or maintenance of economic opportunity throughout the province of Alberta.

Methodology

Satisfaction of the objectives of this paper requires measurement and description of the magnitude of provincial inter-industry linkages for individual Calgary manufacturing firms. This information, in the form of value of provincial purchases and sales of manufactured goods, was obtained from 190 Calgary firms. Each firm also provided an estimate of the extent to which it purchased commodities from, or made shipments to, Alberta manufacturers in six industrial groups. Composition of these six industrial categories was determined by the nature of commodities produced and by the general economic sector with which individual firms were associated. With only two minor exceptions (synthetic textiles and agricultural machinery) the six industrial groups in this analysis are extensions of broad industrial categories and groupings found in the Standard Industrial Classification Code (Appendix I). The estimate by each firm of its linkage with firms in the six industrial groups have been aggregated to describe the industrial linkages of Calgary manufacturers with Alberta manufacturers. Data limitations precluded application of such techniques as interregional input-output or linear programming analyses but the unique character of the data themselves offered insights into the general importance of manufacturing linkages to firms in the Calgary secondary industrial economy which so far have not emanated from any other source.

Published Data

Statistics Canada does not collect locational data related to inter-industry linkages during its annual census of manufacturing, but concentrates instead simply on the measurement of the magnitude of purchases, production, and sales (shipments) of individual firms. Data are published by areal statistical unit provided that they do not violate Statistics Canada's rules of confidentiality. Published data for manufacturers in Canada, however, do not contain information which permit spatial and locational assessment of inter-industry industrial relations. Published data do not permit analysis of individual firms and at best reveal some static locational information about industries or sectors in selected places and regions if the rules of confidentiality are not endangered. The government itself seldom obtains industrial information on origin and destination of materials even for inhouse spatial research; furthermore, the rather voluminous but geographically sterile information contained in individual census returns is not commonly available to non-governmental researchers. Thus suitable flow and linkage data do not exist even in the labyrinth of government, and the data which are collected, although of limited direct spatial value, are restricted in disaggregated form to government personnel.

Data Compilation

The investigator of inter-industry manufacturing linkages must obtain has own information. Most researchers seek data directly from personnel of key firms through such devices as questionnaires. This procedure is expensive, in money and time, both for the researcher and for industrial management. Few firms, if any, seem to maintain adequate information organized in a manner suitable for the geographer. Few manufacturers appear to know the specific origin of all their commodities; they do not maintain files to separate purchases of raw and manufactured goods, and they do not differentiate purchases by "neat" statistical areas such as metropolitan location or province. Many firms purchasing through such middle men as wholesalers are unable to identify either the original producer or even the country or origin of many commodities required. In Alberta, where many manufactured products are produced by assembly of items from diverse suppliers, few manufacturers can with certainty answer in general terms even the simplest questions posed by the industrial geographer if those questions pertain to origin and destination. Many entrepreneurs do not perceive industry in the same manner as the geographer and often refer to construction as an industry, or to retailing as an industry akin to those enterprises engaged in manufacturing. Many firms listed by Statistics Canada as manufacturers plead, when contacted, that they are primarily engaged in services such as repair and maintenance and undertake manufacturing only as an auxiliary and custom activity. In all cases, however, even when the entrepreneur is willing to assist with industrial research, his own files are usually inadequate to provide accurate spatial data.

Most manufacturers also appear to be victims of questionnaire -overkill. They are subjected to questionnaires from various federal, provincial, and local agencies to the point where some claim that they must employ an accountant simply to report data to official organizations. Many claim that the data they have supplied should be released by the government; others are disillusioned with all types of industrial eoncomic research undertaken by government and related institutions because nothing positive seems to emanate from the statistics submitted. One of the biggest obstacles facing the industrial geographer when approaching the entrepreneur thus seems to be the plethora of demands for data made by government.

Spatial inter-industry linkage analyses of the manufacturing economy cannot proceed without the co-operation of industrial management. The data base of the present research (gathered as part of a general analysis of petroleum-generated manufacturing linkages in Alberta) was prepared from a short questionnaire sent to each manufacturer in Calgary who expressed an interest in the project during a preliminary telephone conversation (Appendix II). Since by late 1971, the Alberta Bureau of Statistics had almost completed its Census of Manufacturing for 1968 based on returns from questionnaires submitted to

Alberta firms, the same firms were contacted by the author during the summer of 1972 and asked to supply data for the year, 1968. The return on those questionnaires mailed to entrepreneurs in the city of Calgary was approximately 55 per cent.

Experience by other analysts suggested that most questionnaires prepared by industrial geographers and regional scientists not only ask for too much data, but ask questions which even the most willing respondent cannot answer. Many questionnaires also do not provide the respondent with any incentive to reply or to identify with the objectives of the research or the researchers. The questionnaire utilized in this study sought to overcome these objectives through brevity in the questions asked and through personal contact via the telephone to arouse the interest and sympathy of the person whose co-operation was essential if the questionnaire were to be returned.

The questionnaire sought sixteen pieces of information (Appendix II). Section one asked for the value of all manufactured goods purchased and shipped by the firm, the proportion of total purchases comprised by Alberta-produced Commodities, and the proportion of total shipments which went to customers in Alberta. Section two asked the respondent to allocate the share of manufactured goods either purchased from, or shipped to, Alberta manufacturers according to six industrial categories. The types of industry comprising each category were contained on a sheet attached to the questionnaire. (Appendix II). The entire population of manufacturers were contacted; sampling in the Alberta context could not yield meaningful results due to the small number of manufacturers in many industries. The questionnaire was mailed to each entrepreneur on the same day as he agreed by telephone to assist with the study. The questionnaire was introduced by a covering letter reiterating the objectives of the research and promising a summary of the research findings for each respondent. A self-addressed stamped return envelope accompanied each questionnaire. If the questionnaire were not returned within a month, the potential respondent was again contacted and where necessary was mailed another copy of the questionnaire. In the second phase of the general research programme (not analysed in this paper) those firms which completed the questionnaire were approached in 1973 with a questionnaire designed to measure entrepreneurial attitudes toward the economic milieux of Alberta and Calgary.

Data Utility and Questionnaire Limitations

How useful and reliable are the data obtained for this study? Most entrepreneurs seem to have provided the most recent annual data they had available and not that for 1968. Others responded with data for their own fiscal year. Hence, the years to which responses to questions 1 and 3 pertain vary within the study and may comprise all of, or part of, 1968, 1969, 1970, or 1971, or may extend from 1971 into 1972. The data at the best are probably reasonably accurate and do not vary considerably in quality from information derived from the questionnaires employed in the Census of Manufacturing. The responses to questions 2 and 4, and to all those in part II, clearly are "guesstimates"; no entrepreneur maintains the records or staff necessary to complete these questions precisely. These values must be accepted at face value and treated with great caution; they probably reflect the respondents' perception of the magnitude of his firm's inter-industry linkage as much as the real nature of product movement. Conclusions drawn from this study, therefore, are tentative and must be corroborated by similar studies in which uncertainty and inaccuracy are controlled and reduced. The data limitations will likely plague all such studies in Canada, however, until fundamental changes are made in the way entrepreneurs maintain their records, and government agencies collect and disseminate spatial economic data pertinent to the indivdual firm.

Experience in approaching Calgary manufacturers suggests that there is a definite limit to the quality and depth of information which a researcher can expect to obtain from the entrepreneur or manager. The simple questions asked in the questionnaire heavily taxed the ability of most firms to reply. Refinements to the type of inter-industry locational data which can be obtained probably should be made on the questionnaire forms of the annual census of manufacturing; data provided on purchases and shipments should be accompained by the location of supplier and consumer. However, even this refinement would probably not benefit the independent researcher because it would compound the problem of confidentiality and corporate security. In a general survey such as that conducted for this study, the problem of obtaining inter-industry locational data is probably unresolvable in our society. Some investigators have overcome certain handicaps by concentrating on one industrial sector such as mobile homes or iron and steel where in-depth interviewing can be conducted and where more information is available as part of the general knowledge about the industry in trade literature, etc. The smaller number of firms involved in most specific studies also lessens the cost in time and money involved with obtaining data.

The questionnaire used in the study was purposefully very specific in its objectives. The questionnaire did not seek information on the share of purchased manufactured goods in the total value of all materials purchased by the firm. Although this information would be interesting to analyse in other studies, it would not greatly enhance our understanding of Calgary's inter-industry manufacturing linkages and would probably lessen the willingness of the entrepreneur to provide any information because it would lengthen the questionnaire. Similarly, information on the extent to which commodity shipments are comprised of goods purchased or goods of own manufacture was not requested. The questionnaire also did not ascertain the extent to which the firm was engaged in activities other than manufacturing, or the number or range of commodities produced in any establishment.

The extent to which Calgary manufacturers were linked with other producers in this city was not measured; at the time of the questionnaire, and indeed throughout this entire research project, attention focussed on the need to ascertain provincial patterns of linkages and the extent to which some sort of industrial manufacturing complex exists in the province of Alberta. Further identification and analyses of the internal structure of Calgary (and Edmonton) must be left to the future and to other researchers because of the time and effort involved in such analyses, and because of the plethora of related questions which would have to be asked in a study of intra-urban location.

LINKAGE, EXTERNAL ECONOMIES, AND RELATED STUDIES

General Analyses and Findings

Many of the general problems and achievements related to studies of inter-industry linkage and external economies have been reviewed by such authors as Britton,[5] Karaska,[6] Richter,[7] Streit,[8] Gilmour,[9] and most recently, Czamanski,[10] and Schmidt.[11] Wilbur Thompson[12] has provided a lucid and succinct overview of the processes of internal economic growth generally associated with urban places. Most discussions of linkage and external economies suggest furthermore that shortcomings in statistical data will continue to plague analysts for the foreseeable future. While considerable gains have been made in understanding regional inter-industry linkages, and in appreciating some of the general relationships between linkage and agglomeration or external economies, see for example Beyers,[13] Campbell,[14] Moseley,[15] and Lever,[16] inter-industry linkages at the level of the firm are still relatively few in number and have largely been confined to studies of firms in particular industries, exemplified by the works of Hodge and Wong,[17] Steed,[18] Schmidt,[19] and Taylor,[20] in localities where the existence of manufacturing industries and industrial complexes has long been recognized although perhaps not adequately researched and analysed. Studies of inter-industry linkages at the level of the firm have been confined also to industries whose backward and forward industrial material relations could be reasonably determined from private and public sources. Functional cohesion within industries in processes and products appears to be a prime requisite if meaningful statements are to be made on the influence of size, product mix, ownership, and productivity on material linkage networks among manufacturing enterprises.[21]

Studies, however, of inter-industry material linkages among firms in weakly-developed urban-industrial regions in which the majority of firms purchase materials directly or indirectly from extra-regional suppliers, or in which the bulk of final local demand for industrial commodities is directly supplied by extra-regional producers, are almost entirely missing from the current literature. In other words, studies of manufacturing at the level of the firm have in their recent formative period primarily been concerned with industries and regions in which the magnitude of manufacturing employment and production were important phenomena. The examination of firms' manufacturing linkages in other regions, such as resource-extractive regions with heavy emphasis on the export of unprocessed commodities, has not received adequate attention despite the obvious implications of such research for notions of regional development, diversification, and urban stability. Thompson has observed that, despite the quantification of flows among given sets of local industries in local input-output tables, planners and economic geographers still do not have "the full development equivalent, a locational matrix that tells us which industries follow a given industry to a locality and when".[22] The concomitant of this observation, made by Thompson in a slightly different context, is that examination of the timing of growth, development, and functional expansion is necessary if we are to evaluate "the process by which structural change takes place in the local industry-mix"[23] and to confirm either the development of local agglomeration economies and regional externalities, or the relative growth and dominance of residentiary industries in face of the decline of the original propulsive primary industry.

The diverse and weakly developed manufacturing economy of a city such as Calgary suggests that the search for evidence of strong intra-industry absorption of locally produced manufactured commodities should not be a prime focus of concern; enough is known from existing studies of industrial regions that this pattern of linkage is mainly a characteristic of mature manufacturing complexes.[24] In examining the manufacturing structure of Calgary, located on the periphery of national and international manufacturing complexes, but lacking sufficient local or regional demand to stimulate strong intra-urban linkages of its own, we should instead seek to determine the nature of production chains between Calgary and extra-regional producers, or the extent to which some non-manufacturing regional economic activity has been able to generate limited local manufacturing chains. In essence, we might expect that Calgary and other far-flung Canadian regional centres are able to take advantage of mobile external economies in their backward linkages with large and small extra-regional producers of technologically-intensive products who are spatially accessible to national and international markets, or who depend on access to numerous related products. This expectation would help explain the general absence from Calgary of those manufacturing establishments whose output is characterised by large volumes of standardised commodities, and whose manufacturing processes depend on the mass assembly of numerous manufactured sub-components.

Cities such as Calgary, however, should be expected to display reasonably stong manufacturing linkages among producers of perishable agricultural commodities prior to shipment to final regional and extra-regional markets, and to display significant linkages among firms related to the production of custom goods or small-run items related to resource-development activity, and industrial and residential construction. In this manner, local and regional manufacturers are able to take advantage of agglomeration economies in Calgary but rely for many standardized products on manufacturers in extra-regional industrial complexes such as the Quebec city-Windsor industrial corridor.

The nature of urban manufacturing linkages with adjacent urbanized and non-urbanized space clearly depends on the nature of the regional economy. The past quarter century in Calgary's region has been associated with the rapid development of petroleum extraction for extra-regional export. This activity not only has prompted the expansion of tertiary and quaternary industries in Calgary (discussed elsewhere in this volume) but also has led to the demand for development-related equipment. Those Calgary firms which have been able to operate at sufficient scales to compete with extra-regional producers have been able to satisfy important aspects of regional demand for development equipment. Some Calgary firms have added relatively little value to manufactured goods imported from extra-regional manufacturers; others have developed sophisticated production lines and generated important, although limited, inter and intra-industry manufacturing linkages.

The city's relationship with regional manufacturing can be identified through the provincial linkages of its manufacturing firms tentatively explained in relation to many of the findings and concepts of urban-economic and inter-industry linkage analyses in other world and national regions. In this vein, the present study owes considerable debt to those who have helped develop a strong and fertile body of literature in this field; the empirical findings for Calgary presented in this paper attempt to add to the general understanding of the function of inter-industry manufacturing linkages not only in this city but also in all intermediate-size urban places which lie spatially beyond the bounds of major industrial complexes.

Analyses of Alberta

A recent review of studies of the Alberta economy contains few references to other analyses of material flows among sectors of the provincial economy.[25] One major input-output analysis has been completed for the province using 1962 data.[26] Unfortunately, however, urban input-output relations within the provincial economy were not the object of Wright's analysis and cannot be inferred from his provincial patterns. However, Wright's study does add a perspective to the present analysis of Calgary's manufacturing linkages by demonstrating the close similarity between the impact of petroleum and agriculture on the provincial economy; two of the strongest sectors of the Alberta manufacturing economy relate to the processing of agricultural commodities, and to the production of primary metals, engineering, and transportation equipment for eventual consumption by the petroleum industry or by activities generated from the growth of the primary petroleum industry.[27]

Other Alberta analysts, most notably E.J. Hanson,[28] have devoted considerable effort to analyses of the employment and income effects of the Alberta petroleum industry, and have shown the dependence of a major share of provincial employment directly or indirectly on exploration and development in the petroleum industry. Hanson's work did not concern itself specificially with the impact of resource-extraction on urban manufacturing complexes but did provide a necessary perspective within which to identify those economic processes associated with the general provincial expansion of this propulsive industry since 1947.

Few analyses of the spatial impact of petroleum on the Alberta regional and urban economy have been carried out. Entries in a recent review of urban and regional studies in Alberta for the period 1950–1974 suggest that the impact of petroleum (the dominant growth sector in the provincial economy throughout this period) on the spatial structure of the economy has either been considered unworthy of investigation, or so minor as not to warrant serious study. With few exceptions, geographers and economists alike have ignored the spatial structure of Alberta's resource-based economy. Drugge[29] has specifically examined some of the locational requirements of energy-intensive industries in Alberta. Peter Smith's[30] work on Fort Saskatchewan and Calgary stands alone in its concern for the changing industrial composition of two important urban economies. The changing provincial distribution of manufacturing has been analysed in short papers by Leigh and Carter,[31] and Seifried;[32] Barr[33] has considered Alberta's industrial diversification as part of his analyses of post-World War II economic change in the three prairie provinces. Theses on the subject of petroleum-generated industrial location have been confined to analyses by Zieber[34] of the location of oil offices in Edmonton and Calgary, by Curtis[35] of the generation of tertiary industrial linkages in Edmonton with the petroleum industry, and by Wilson[36] of some of the factors attracting manufacturing to Alberta.

The rather weakly developed portfolio of existing analyses and the narrow profile of their interests have led Barr and Fairbairn[37] to embark upon a comprehensive analysis of petroleum-generated manufacturing linkages within Alberta. This report on Calgary draws heavily on data obtained by the author as part of his wider examination of linkages within the Albertan economy.

READ REST OF CHAPTER AS GO ALONG PROBABLY DISCUSSION IS ESSENTIAL

MANUFACTURING IN CALGARY

Calgary's Manufacturing Structure

Over two-fifths of all manufacturing establishments in Calgary are engaged in general manufacturing; an additional one-fifth of Calgary's firms are associated with primary metal engineering, or transportation equipment, and another fifth are engaged in the production of foods and beverages. When measured according to the criterion, selling value of factory shipments of goods of own manufacture, however, the importance of general manufacturing and foods and beverages is almost reversed. Primary metals, engineering, and transportation equipment industries comprise one-fifth of the Calgary economy (Table 1,1). This share does not vary among the three measurement criteria used in Table 1,1. However, the considerable disparity between the relative importtance of food and beverages, and general manufacturing, according to the criterion employed seems to reflect the prevailing large size of entreprises in the food and beverage industry compared to those in general manufacturing. This difference in size of establishment further suggests that the general function of production of foods and beverages is to satisfy a significant share of the Canadian market for those commodities whereas many general and non-food manufacturers appear to be providing a wide-range of goods to a local and regional economy.

The structure of Calgary's manufacturing economy clearly does not display a strong association with the processing or conversion of petroleum. The city's image as petroleum capital cannot be found reflected in its manufacturing structure. In this regard, Calgary's manufacturing profile is almost identical with that of Alberta--neither entity has a strong association with the conversion of petroleum into manufactured goods. Edmonton, however, while almost identical in manufacturing structure to Calgary and the province, does have a larger

TABLE 1,1

STRUCTURE OF MANUFACTURING - 1970

INDUSTRIAL GROUP	ALBERTA						CALGARY						EDMONTON					
	ESTABLISH-MENTS		VALUE ADDED		SHIPMENTS		ESTABLISH-MENTS		VALUE ADDED		SHIPMENTS		ESTABLISH-MENTS		VALUE ADDED		SHIPMENTS	
	NO.	%	$\$ \times 10^6$	%	$\$ \times 10^6$	%	NO.	%	$\$ \times 10^6$	%	$\$ \times 10^6$	%	NO.	%	$\$ \times 10^6$	%	$\$ \times 10^6$	%
1. Food & Beverage	463	24	185.	27	764.	40	87	17	57.	26	201.	37	93	16	74.	30	282.	39
2. Petrochemical, Coal and Synthetic Textile	93	5	114.	17	318.	17	35	7	30.	14	75.	14	44	8	35.	14	171.	24
3. Non-Food, and General Manufacturing	592	30	117.	17	204.	11	213	42	43.	20	77.	15	226	39	54.	21	93.	13
4. Wood, Paper, and Allied Industries	374	19	77.	11	168	9	44	8	23.	11	51.	9	48	8	12.	5	28.	4
5. Primary Metal, Engineering, Transportation Industries	324	16	134	19	347	18	112	22	51.	23	113.	21	139	24	42.	17	92.	13
6. Non-Metallic Mineral Products	110	6	63.	9	102	5	20	4	12.	6	21	4	32	5	33	13	52.	7
TOTAL	1956	100	690	100	1903.	100	511	100	216.	100	538	100	582	100	250	100	718.	100

Source: Industry and Resources 1973. Alberta Bureau of Statistics, Alberta Industry and Commerce, 1973, pp. 25-33; missing data have been estimated from other sources.

share of its manufacturing structure related to petrochemicals, especially in terms of the criterion, shipments of goods of own manufacture. The stronger showing by the petrochemical industry in terms of shipments instead of value added suggests that Edmonton's petrochemical industry, while contributing to the manufacturing profile of that urban region, does not engage in significant downstream processing of petroleum and the generation of major multiplier effects. If there were a strong interdependence and production chain related to petrochemicals in the Edmonton region, we should expect to see that industry contribute a much greater share to the industrial structure when measured in terms of value added.

Although Calgary is weaker than Edmonton in petrochemical industries, it appears to have a greater strength in primary metals, engineering, and transportation equipment. Calgary in 1970 exceeded Edmonton in the magnitude of these industries although Edmonton had a greater number of establishments in this general industrial category. Calgary's superior strength in these industries suggests that this city has a stronger relationship with oilfield developers in terms of metals, pipe, control structures, and transportation equipment. This advantage probably stems from the initial development of the petroleum industry in southern Alberta. However, these observations are general and await considerable research and verification before substantial conclusions can be drawn with any degree of confidence.

The contribution of Calgary and Edmonton to the provincial manufacturing economy is so great that they seem in fact largely to determine the general structure of the aggregate provincial economy. Edmonton does show a more advantageous locational relationship with the petroleum resources and will likely enhance its share of the petrochemical industry in succeeding years as the petrochemical industry continues to rationalize a considerable amount of its prairie capacity in the Edmonton region; Calgary's primary metal, engineering, and transportation equipment industries deserve careful monitoring in subsequent years as the locus of petroleum development continues to move northward into arctic frontier regions.

Provincial Commodity Shipments

A comprehensive perspective on provincial inter-sectoral flows of commodities is provided by Wright's 1963 input-output analysis.[38] His study analyses all economic flows through the provincial economy, and also estimates provincial export and import of commodities. Wright's report, furthermore, adds a perspective to the present study by presenting manufacturing flows as part of the aggregate contribution of primary and secondary industry, and the "service industries" (which obviously include tertiary and quaternary functions). Flows among the manufacturing sectors in his transaction matrix, technological matrix, and inverse matrix can be reworked according to the six industrial groups employed in the present analysis; Wright's inter-industry industrial linkages within the Alberta manufacturing economy differ in most cases from those observed here for either the Edmonton or Calgary urban economies but final comparison with total Alberta linkages will have to await complete processing of the larger study by the present author and K.J. Fairbairn. A major problem in completely adapting the manufacturing flows in Wright's matrixes to the needs of the industrial geographer in Alberta arises in attempts to distinguish the consumption of imports of Alberta manufacturers from those by other provincial consumers; similarly, the extent to which exports arise with the manufacturing economy as compared to the other provincial producers cannot be determined from his matrix. However, Wright's intra-provincial flows can serve as a norm against which to measure some of the empirical findings of this research, and against which to estimate some of the changes which have occurred in the provincial

PLATE 5
Calgary, circa 1926.

PLATE 6
Calgary, 1945.

Glenbow Foundation Archive Photo

economy between 1962, when petroleum exploration and development were dynamic, and 1972, when a major thrust of the Canadian search for petroleum had already shifted to the north.

Calgary's manufacturing structure suggests that this city should be summarized and described in relation to secondary industry as a diversified general manufacturer related to three major aspects of the regional and provincial economy--the processing of regional agricultural commodities for regional and national consumption, the production of oilfield equipment for the provincial and northern primary petroleum industry, and satisfaction of the demand by the regional population for a wide range of light industrial commodities whose high friction of distance protects specific Calgary manufacturers from extra-regional competition.

This characterization of Calgary's manufacturing economy, however, must be supported by evidence which does indeed carry with it some indication of functional connectivity. Aggregate data on the destination of shipments, however, are not available for Calgary's secondary sector and, indeed, have only recently been published for the province of Alberta. (Table 2, 1). If we conclude, therefore, that Calgary's industrial structure and hence its function closely corresponds to that of the province, then aggregate provincial data for 1967 might be employed to infer the extent to which Calgary's major manufacturing activities do in fact carry with them the functional subordination and role ascribed to them in the previous paragraph.

The Alberta food and beverage industry ships just over half of its goods to consumers in Alberta and satisfies two-thirds of the demand for these products in the province. The Alberta, and hence, Calgary food and beverage industries therefore have major extra-regional markets which suggest that this industrial group has strong forward linkages with consumers outside of Alberta. This same general relationship holds for the petrochemical and wood and paper industries. Clearly these man-

TABLE 2,1
SHIPMENTS OF GOODS OF OWN MANUFACTURE (1967)

INDUSTRIAL GROUP	SHIPMENTS BY ALBERTA MANUFACTURERS:				SHIPMENTS TO ALBERTA BY ALL CANADIAN MANUFACTURERS
	(1) TO ALBERTA			(2) TOTAL TO ALL CUSTOMERS*	
	$\$ \times 10^6$	(1) as % of (2)	(1) as % of (3)	$\$ \times 10^6$	$\$ \times 10^6$
1. FOOD & BEVERAGES	345.	54	67	642.	512.
2. PETROCHEMICAL, COAL, & SYNTHETIC TEXTILE	156.	55	49	282.	318.
3. NON-FOOD, & GENERAL MANUFACTURING	85.	65	19	132.	439.
4. WOOD, PAPER, & ALLIED INDUSTRIES	67.	48	45	139.	151.
5. PRIMARY METAL, ENGINEERING, TRANSPORTATION INDUSTRIES	159.	66	34	240.	463.
6. NON-METALLIC MINERAL PRODUCTS	72.	78	85	92.	85.
TOTAL	884.	58	45	1,527.	1,968.

Source: "Destination of Total Shipments of Goods of Own Manufacture, 1967," Destination of Shipments of Manufacturers, Catalogue 31-504 Occasional, Information Canada 1971, Table 2; Industry and Resources 1973. Alberta Bureau of Statistics, Alberta Industry and Commerce, 1973, p. 18. Missing data have been estimated.

* Includes customers in foreign countries.

ufacturers located in Alberta, while not producing the full range of goods necessary to supply the total regional demand for these products, do nevertheless have strong regional and extra-regional forward linkages. The data do not, however, indicate if these shipments were made to other manufacturers or to consumers in other sectors of the economy. Disaggregated spatial flow data are not available from government and for this study must be obtained through questionnaire surveys.

Nearly two-thirds of the output of the primary metal, engineering, and transportation equipment industries is destined for provincial consumption; surprisingly, however, Alberta manufacturers supply only one-third of the total provincial demand for these manufactured goods. Although extra-regional supply of automobiles probably reduces the significance of the Alberta manufacturers to total consumption of goods in this group, the export destinations of most commodities manufactured by industries in this industrial group is chiefly confined to adjacent Canadian political regions.

Alberta manufacturers engaged in non-food and general manufacturing, and production of non-metallic mineral products ship the majority of their commodities to provincial consumers; these same manufacturers, however, vary in the extent to which they satisfy the total provincial demand for commodities produced by firms in their industrial groups. Producers in the non-metallic mineral industries seem to enjoy a protected market due to the generally high friction of distance associated with their commodities, the provincial availability of necessary raw materials, and the ability of regional demand to support economic enterprises and scales of output in these industries. General and non-food manufacturers, however, while shipping a major share of their product to provincial consumers satisfy less than one-fifth of the total provincial demand for commodities produced by this industrial group.

This dichotomy-- while not the focus of this paper--probably is related to such factors as size of local market, general economies of scale--both internal and external--operating within the industries of this group, and general comparative advantages held by producers in those major Canadian and foreign industrial regions which ship goods to Alberta.

CALGARY'S AGGREGATE PROVINCIAL MANUFACTURING LINKAGES

Previous sections of this analysis rely heavily on inferences from published aggregate production and shipment data to determine Alberta's significance as a manufacturing region. One of the major questions which cannot be answered from these data or from other published data is the extent to which manufacturers in Calgary are part of a provincial manufacturing complex. Two major types of evidence must be presented before Calgary's functional role with other provincial manufacturers can be ascertained: the extent to which Calgary manufacturers purchase manufactured goods within the province of Alberta, and the extent to which Calgary producers ship manufactured goods to other Alberta manufacturers.

The questionnaire responses permit, for the first time, meaningful although tentative statements to be made about many aspects of Calgary's manufacturing linkages. The value of shipments made by respondents in each industrial group comprise a significant share of the total value of shipment by these groups published by Statistics Canada. Clearly the size of industrial production of those who responded to the questionnaire is sufficient to expect the patterns of industrial linkage reasonably to reflect patterns for all manufacturers in Calgary.

Backward Linkages

The strongest backward linkages are evidenced by food and beverage industries, and by non-metallic mineral producers (Table 3,1). Producers in these industries rely heavily on semi-processed goods and on

TABLE 3,1

AGGREGATE INDUSTRIAL LINKAGES OF CALGARY MANUFACTURING PLANTS

INDUSTRIAL GROUP	PURCHASE OF MANUFACTURED MATERIALS & SUPPLIES:					SELLING VALUE OF FACTORY SHIPMENTS:				
	TOTAL		FROM ALBERTA MANUFACTURERS:		ALBERTA PURCHASES	TOTAL		WITHIN ALBERTA:		ALBERTA SHIPMENTS
	NO. OF ESTABS.	VALUE $\$ \times 10^6$	NO. OF ESTABS.	VALUE $\$ \times 10^6$	AS % OF TOTAL	NO. OF ESTABS.	VALUE $\$ \times 10^6$	NO. OF ESTABS.	VALUE $\$ \times 10^6$	AS % OF TOTAL
1. FOOD & BEVERAGE	23	6.3	23	4.0	64	23	84.6	23	28.7	34
2. PETROCHEMICAL, COAL, & SYNTHETIC TEXTILE	8	5.1	7	2.0	39	8	43.6	8	13.2	30
3. NON-FOOD, & GENERAL MANUFACTURING	78	18.8	58	1.9	10	78	37.2	76	26.3	71
4. WOOD, PAPER, & ALLIED INDUSTRIES	15	15.2	13	2.4	16	15	24.8	15	14.9	60
5. PRIMARY METAL, ENGINEERING, TRANSPORTATION INDUSTRIES	52	60.2	50	11.7	19	52	97.4	51	62.9	65
6. NON-METALLIC MINERAL PRODUCTS	14	9.2	13	6.2	67	14	18.1	14	17.7	98
TOTAL	190	114.8	164	28.2	25	190	305.7	187	163.7	54

Source: Questionnaire survey, May - August, 1972.

packaging materials in the form of paper, paperboard, plastics, glass, etc. Calgary's petrochemical industry similarly makes significant local purchases of packaging materials but the relatively weak backward manufacturing linkages appear to reflect its purchase of chemical bases and related materials from producers outside of Alberta.

The remaining three groups exhibit very weak backward manufacturing linkages within the province. Non-food and general manufacturing industries clearly rely on sub-components produced in central Canada and foreign industrial regions. These manufacturers are clearly at the end of the production chain and are assembling manufactured goods for sale to the final consumer. Alberta's economy in this respect must clearly be described as consumer-oriented. Wood, paper and allied industries also rely very heavily on lumber and paper from outside the province. Glues and inks originate from elsewhere. Alberta does not have a finished paper industry and at the time of our survey had only one pulp mill, the primary product of which is shipped entirely outside the province for conversion into paper. The major Alberta purchases of manufactured goods by these Calgary industries are probably for commodities which do not directly enter into the production process or comprise a real share of the final product. Enterprises engaged in primary metals, engineering, and transportation also purchase a major share of their manufactured goods outside the province but the supply of components to the engineering and transportation industries probably means that goods purchased within the province do enter into the final product of these firms and are not adjunct to the main manufacturing process. However, the backward linkages of groups 3 - 5 reveal how dependent Calgary truly is on manufacturing processes and enterprises in other parts of Canada or the world.

Although the proportion of manufactured goods purchased within Alberta varies from one group to another, most establishments in

Calgary appear to make some purchase of manufactured goods within Alberta. In all groups except non-food and general manufacturing, nearly all respondents to the survey reported making some purchase within Alberta. Non-food and general manufacturing firms have the weakest backward links with provincial manufacturers, and many of them make no significant purchases at all of goods manufactured within Alberta.

Forward Linkages

While most respondents were able to identify the extent to which they purchased goods manufactured within Alberta, many were unable to determine the extent to which manufacturers, as opposed to all consumers, consumed their products because so many of their commodities were marketed through wholesalers and agencies. The questionnaire sought, therefore, to determine the extent to which sales were made to any customers in Alberta. Consequently, the data in Table 3, 1 appear at first glance to suggest that the linkage between Calgary manufacturers and consumers in Alberta--forward linkages--involve more firms than those associated with backward linkages. All but three respondents reported making some shipment to Alberta customers but, when data from part II of the questionnaire were carefully analysed for evidence of shipments to Alberta manufacturers, only 97 out of 190 respondents provided evidence that they shipped to manufacturers. Thus, the number of firms reporting backward linkages (164) is far greater than the number making shipments to Alberta manufacturers. Many Calgary manufacturers are at the tail-end of the production chain; nearly half of the firms surveyed were engaged in production of goods for final consumption, either by the public or various non-manufacturing sectors of the provincial economy.

Four Calgary industrial groups have linkage relations with

provincial consumers which are different from those observed for all provincial industries in the same groups. Calgary's food and beverage industries ship only one-third of their production to Alberta consumers compared to just over half shipped to provincial consumers by all Albertan food and beverage industries. The same relationship prevails for manufacturers in the petrochemical group. For these two Calgary groups, markets lie mainly outside the province and are spatially quite different from those of the other provincial manufacturers of these commodities. Wood and paper industries in Calgary might correspond in shipment pattern to the province as a whole if shipments from the Hinton pulp mill were deleted from the provincial figures. Calgary's non-metallic mineral products industries, however, are almost entirely linked with provincial consumers and appear to have a much greater friction of distance associated with their commodities than many of the manufacturers of commodities in this group in the rest of Alberta. Calgary's regional ties with consumers of commodities produced by non-food and general manufacturers, and by primary metals, engineering, and transportation equipment enterprises are almost identical to those of all Alberta industries.

Linkages and Functional Diversity

Calgary is thus less dependent on Alberta markets for food and petrochemical products than other Alberta manufacturers, but is more dependent than other provincial manufacturers on Alberta demand for non-metallic mineral products. Even more significant than this observation, however, is the discovery that nearly half of Calgary's manufacturers do not sell to manufacturers in Alberta but instead produce for the final consumer. On balance, therefore, Calgary manufacturers have a greater number of backward linkages than forward linkages with Alberta manufacturers; the realization that 67 more Calgary manufacturers bought from Alberta manufacturers than sold to them suggests that not only is Calgary producing for the final consumer, but that this city's manufacturers are not primarily engaged in buying from each other.

Calgary's manufacturing economy relies for only one-quarter of its purchase of manufactured goods from Alberta producers; this relationship suggests that Calgary's manufacturing economy is really part of manufacturing complexes or systems in other regions of Canada or the world. Calgary producers are largely at the end of the production chain but the initial links of that chain are not found primarily in Alberta. On the other hand, over half of Calgary's factory shipments are to consumers in Alberta; and approximately half of Calgary's shipments to Alberta customers are destined for Alberta manufacturers.

The sales of Calgary-manufactured goods to Alberta manufacturers, however, is approximately three times greater than Calgary's purchases of Alberta manufactured goods. This observation suggests that the number of linkages of Calgary firms with other Alberta manufacturers is high but the average dollar value of purchases is low; fewer Calgary firms ship to Alberta manufacturers but the average value of their shipments is high. Although Calgary's relatively weak linkages with other Alberta manufacturers does not seem unexpected, the city's general reliance on extra-provincial sales of its manufactured goods for approximately half of all sales of these goods prevents the manufacturing economy of this city from being unequivocally labelled as a regionally oriented producer of manufactured goods. Although Calgary's overall economic role might show strong signs of regional provision of wholesale commodities and petroleum services, its manufacturing sector due to food and beverages, petrochemicals, and primary metals, engineering, and transportation equipment industries, is far from being primarily dependent on regional or provincial markets.

CALGARY'S PROVINCIAL INTER-INDUSTRY LINKAGES

Backward Linkages

In their purchases of Alberta manufacturered goods, Calgary firms are most closely linked with the primary metal, engineering, and transportation equipment industries; approximately two-fifths of Calgary's purchase of provincially-manufactured goods is from firms in this group while enterprises in the non-metallic mineral industry supply about another fifth. Calgary's heavy reliance on metal and fabricating industries is consistent with the provincial demand for oilfield equipment and with the general provincial economic reliance on primary industry. The manufacturing economy of Calgary bears almost no direct relation to the manufacture of petroleum. Only as a relatively small supplier of total provincial demand for manufactured products by the primary petroleum sector has, Calgary, like Edmonton, derived some benefit to its secondary industry. The linkage of Edmonton manufacturers with producers of petroleum products is almost as weak as that of Calgary producers; returns from the questionnaire survey in Edmonton showed that 7.9 per cent of respondents' purchases of Alberta manufactured goods was from the petrochemical industry. This compares to a figure of 4.3 per cent reported by Calgary respondents (Table 4,1).

Perhaps it is unreasonable to expect the Alberta petrochemical industry to play a large role as a supplier to provincial manufacturers given the small regional population and the great distances to North American and world industrial and consumer markets. Locational inertia caused by such factors as existing population and industrial distribution in Canada and the "headstart" in manufacturing possessed by producers in Ontario and Quebec obviously must have some great bearing on the relative underdevelopment of petroleum-generated manufacturing linkages in Alberta and in Calgary. While we have not measured the

TABLE 4,1

INTER-GROUP PURCHASES OF MANUFACTURED COMMODITIES WITHIN ALBERTA BY CALGARY MANUFACTURERS
(Percentage of Total Purchases, by Each Group, of Commodities Manufactured Within Alberta)

INDUSTRIAL GROUP — BY ↓ / FROM →	1	2	3	4	5	6	TOTAL PURCHASES ($x10^6)
1	75.0	.6	8.0	13.2	.7	2.5	4.0
2	2.5	32.2	6.6	25.1	28.0	5.6	2.0
3	N	8.6	23.0	22.8	44.9	.7	1.9
4	----	----	11.8	71.1	14.2	2.9	2.4
5	N	1.3	9.4	6.7	77.5	5.1	11.7
6	----	4.0	4.2	1.4	18.6	71.8	6.2
PURCHASES BY ALL GROUPS	10.9	4.3	9.0	14.4	42.5	18.9	28.2

N= NEGLIGIBLE

TABLE 5,1

NUMBER OF CALGARY MANUFACTURING ESTABLISHMENTS ENGAGED IN INTER-GROUP PURCHASES OF COMMODITIES MANUFACTURED WITHIN ALBERTA
(Number of Establishments Analyzed: 190)

INDUSTRIAL GROUP — BY ↓ / FROM →	1	2	3	4	5	6	NUMBER OF ESTABLISHMENTS MAKING SOME ALBERTA INTER-GROUP PURCHASE
1	17	3	8	11	5	3	23
2	1	7	4	3	2	4	7
3	1	14	33	39	24	8	58
4	-	-	6	12	4	2	13
5	1	6	27	18	35	7	50
6	-	2	2	4	8	12	13
TOTAL	20	32	80	87	78	36	164

Source: Questionnaire survey, May - August, 1972.

significance of petroleum as a raw material to oil refineries, for example, or natural gas to fertilizer complexes, and cannot comment, therefore, on the extent to which petroleum is converted by primary manufacturing establishments into relatively unsophisticated products such as gasoline and fertilizer, the linkage patterns in Table 4,1 do show that few Calgary firms have any meaningful backward linkages with the petrochemical industry. One of the obvious implications of this observation is that growth, decline or even demise of the conventional provincial petroleum industry will not directly affect to any sizeable extent the pattern of backward material linkages by Calgary manufacturers although curtailment of demand for oilfield and related construction and transport equipment would affect industries in Group 5.

Inter-group linkages often imply that all members of the group behave more or less in the same manner. However, the number of establishments engaged in inter-group purchases clearly is not consistent within all groups. The data in Table 5,1 suggests that some firms in any group probably have a much greater connection with Alberta producers than that of other firms in the group. This observation further suggests that the necessity to group Calgary manufacturers into only six groups has probably masked some important variations which exist between individual enterprises and firms. The groups, while logically valid to the industrial geographer, do not seem to be internally homegeneous in terms of their linkages with other Alberta firms. Many Calgary enterprises are small and many belong to firms with which have plants in other regions of Canada or in other countries. Diverse pattern of ownership, control, functional subordination, as well as important intra-group product differences, appear to add to the heterogeneous nature of Calgary's manufacturing linkages. And the relatively small number of enterprises in each S.I.C. category compounds the difficulty of identifying clear and consistent patterns for aggregations such as the six industrial groupings in Table 5,1.

Any projection or recommendation for industrial policy or strategy based on patterns derived for these six industrial groups for example, must only proceed after careful scrutiny of the internal consistency of each group. This observation also suggests that consistent approaches in industrial development strategy in such a province as Alberta where there are relatively few firms in each industrial category must either be so general as to be specifically ineffective, or must deal with each firm on the basis of its own merits. On the other hand, the diversity of manufacturing in Alberta, Calgary, and Edmonton suggests that policies and programs in the national interest should be pursued in which the welfare of individual enterprises or firms in any region could be enhanced by simultaneous solution of the problems of all members of each industry in the small national Canadian economy. The diversity in linkage patterns in Table 5,1 suggests furthermore that specific measures to aid such manufacturers in Calgary, if repeated for other Canadian urban places by pertinent provincial governments, could adversely affect national industrial performance.

Forward Linkages

Although Calgary manufacturers generally have their strongest provincial backward linkages with firms in their own industrial group, their strongest forward provincial linkages are almost entirely with firms outside their own industrial group. Furthermore, with the exception of primary metal, engineering, and transportation equipment industries, Calgary manufacturers have the strongest component of their provincial sales to non-manufacturers. The patterns in Table 6,1 enhance the observation that Calgary manufacturers are indeed producing for the final consumer (industrial or personal) and are engaged in producing a diverse range of commodities for provincial manufacturers. The forward manufacturing linkage of food and beverage industries is quite clearly with producers of commodities such as agricultural feedstocks and livestock products, and semi-finished food and animal products for further processing prior to shipment to the final individual consumer. Forward industrial linkage of the petrochemical industry is chiefly with the primary metal, engineering, and transportation equipment industries and satisfies their demand for such commodities as petroleum, lubricants and chemicals. Non-food and general manufacturing industries in this study are so diverse in function and product mix that, they as a group, satisfy some demand in every industrial group although their usual function is to produce items for final consumption by the individual consumer. Many products of this industrial group, therefore, might be seen as performing some auxiliary function in manufacturing enterprises rather than as entering the composition of the commodities produced.

Output of the Wood, paper and allied industry group clearly consists of packaging materials and paper and wood products which are consumed in a wide range of industrial and operational functions. This industrial group does perform an important service for the food and beverage industry in the province and supplies a diverse range of products to the general and non-food light manufacturers of Alberta. Industrial group five, primary metals, engineering, and transportation equipment, has the strongest linkage with other Alberta manufacturers of all groups shown in Table 6,1. This group may be thought of in the Calgary context as the most industrial of all in that most of its commodities do not go directly to the final consumer without further manufacture. This industrial group offers the strongest evidence of all groups that Calgary does have at least one manufacturing component which feeds the processing lines of other manufacturers and creates industrial, as opposed to service, multipliers within the province. If one of the criteria for recognizing an industrial complex is that most of the output should go to other manufacturers and not to final demand, then in terms of forward linkage, the only group offering unequivocal evidence of some intra-provincial industrial complex associated with Calgary's manufacturing output is that associated with primary metals, engineering and transportation equipment. The total sales of this group are so large that they dominate those of any other group and, because of their prime orientation toward provincial manufacturers, reduce the total value of the city's reliance on sales to non-manufacturers to just less than 50%. The non-metallic mineral products industry stands in complete contrast to that of group five because the largest share of its output is directed toward the final consumer.

Calgary's forward linages like its backward ones are diverse within each industrial group (Table 7,1). Establishments in each group generally ship some commodity to firms throughout the range of groups; but all groups are similar in terms of the large number of firms which ship to non-manufacturers in the province. The only group where shipments to manufacturers clearly outweigh shipments to non-manufacturers in any major way (the number of firms in the petrochemical group is very small although the ratio of shipments to manufacturers compared to those of non-manufacturers is high) is associated with primary metals, engineering, and transportation equipment.

Although the value of linkages among firms may not be particularly outstanding in terms of major industrial regions, the number of enterprises engaged in some form of inter-industry shipment attests to the widespread influence of inter-industry sales among Calgary's manufacturers. Although Calgary is spatially separated from major industrial regions of North America, and although its aggregate contribution to national value added through manufacturing is very low, inter-industry linages do nevertheless play an important role in this urban economy; urban outliers such as Calgary may not be so industrially unsophisticated as their general role in regional administration and supply usually suggest. Perhaps Calgary's long distance from other industrial

TABLE 6, 1

INTER-GROUP ALBERTA SALES OF MANUFACTURED COMMODITES BY CALGARY MANUFACTURERS
(Percentage of Total Sales, by Each Group, to ALL Customers Within Alberta)

INDUSTRIAL GROUP — BY ↓ / TO →	1	2	3	4	5	6	To Non-Manufacturers	TOTAL SALES ($x10⁶)
1	45.2	.1	.8	.1	---	---	53.8	28.7
2	---	10.1	6.8	1.7	24.8	---	56.6	13.2
3	1.4	16.0	6.8	5.3	5.1	1.1	64.3	26.3
4	27.5	1.0	13.9	1.0	1.1	1.2	54.3	14.9
5	1.2	35.1	9.5	.6	17.9	3.9	31.8	62.9
6	1.1	1.1	4.3	1.1	19.4	4.6	68.4	17.7
SALES BY ALL GROUPS	11.2	17.1	7.2	1.4	11.9	2.3	48.9	163.7

TABLE 7, 1

NUMBER OF CALGARY MANUFACTURING ESTABLISHMENTS ENGAGED IN INTER-GROUP SHIPMENTS TO ALBERTA MANUFACTURERS
(Number of Establishments Analyzed: 190)

INDUSTRIAL GROUP — BY ↓ / TO →	1	2	3	4	5	6	To Non-Manufacturers	NUMBER OF ESTABLISHMENTS MAKING SOME INTER-GROUP SHIPMENTS
1	11	1	3	1	-	-	11	12
2	-	2	4	2	2	-	2	6
3	19	20	30	17	23	12	40	35
4	4	2	3	3	2	2	10	5
5	7	11	20	6	21	8	20	35
6	1	1	3	1	2	3	10	4
TOTAL	42	37	63	30	50	25	93	97

Source: Questionnaire survey, May - August, 1972.

areas coupled with the demand by the Alberta economy for light engineering commodities act to increase the extent of the city's inter-industrial linkage beyond that which would occur if major suppliers of these industrial commodities were geographically closer to Alberta. The specific nature of Calgary's inter-industry linkages is beyond examination in this report but clearly, despite the diversity in goods supplied to the provincial economy, one cannot help but wonder if most of the Calgary industries supplying Alberta industrial and consumer markets do have in common such factors as high friction of distance on their products, or markets in which custom (as distinct from standard) products are demanded.

DISCUSSION

Linkages and the Calgary Economy

What do Calgary's backward and forward industrial linkages with otherAlberta manufacturers reveal about the city's aggregate secondary industrial economy? Approximately one-quarter of Calgary's purchases and shipments of manufactured goods is associated with industries in the province of Alberta; the magnitude of Calgary's inter-industry linkages with the Alberta economy, however, is heavily biased toward forward linkages. Calgary's industrial linkages with the provincial economy attest to the city's relatively minor role as a purchaser of value previously added to Alberta commodities, and major role as an exporter of industrial wealth and utility to the region.

Calgary as a manufacturing system, however, is not primarily dependent on sales within Alberta. Approximately half of its manufactured output leaves the province although it is not possible to specify the extent to which Calgary manufacturers supply extra-provincial manufacturers as distinct from final consumers. Calgary cannot, therefore, be described in terms of its manufacturing function simply as a regional producer.

Calgary's manufacturing economy is diverse; several of its main functions relate to the food and beverage industry and to some of the lighter aspects of primary metals, engineering, and transportation equipment. These functions attest to the city's role in the processing agricultural commodities for regional and national distribution, and in supplying important segments of the provincial and northern demand for oilfield equipment, construction commodities, and other fabricated items.

Calgary's inter-industry linkages are not uniform; this suggests that some sectors of its urban manufacturing economy are provincially much less integrated than others. Those industrial groups which assemble diverse processed commodities still rely heavily on extra-provincial purchases and suppliers. Those Calgary industries not associated with manufacture of food and beverages, or non-metallic mineral products, suggest by the nature of their backward linkages that much of the city's secondary industrial base is related to the assembly and fabrication of products from subcomponents supplied by extra-regional suppliers. This heavy reliance on extra-regional suppliers of industrial subcomponents, and manufactured items in general, attests to the relatively small size of Calgary's urban and regional population, and to many structural characteristics of the Canadian economy which are frequently cited in explanation of regional economic imbalance in the country.

An important segment of Calgary's manufacturing economy is related to the final stages of various national and international production chains. The city obviously has numerous industrial relations with other urban-industrial places. While not the focus of this particular study examination of Calgary's extra-provincial inter-urban economic linkages would add further to our understanding of the purpose and function of the city's linkages. The general effect of friction of distance and extra-regional agglomeration economies on the structure of Calgary's manuufacturing economy is not clear and can only be approached in very loose terms at the moment. Similarly, the extent to which general urban agglomeration economies or economies of scale are operating within Calgary cannot yet be meaningfully specified.

Calgary at the present time is part of a weakly-developed provincial industrial complex. Provincial inter-industry linkages with Calgary firms are reasonably numerous but economically rather weak. In the assembly and light manufacturing industries, supply ties with extra-regional producers are much stronger than those within the province. For many groups one can probably identify the beginnings, or at least the existance , of a weak industrial complex but one cannot with any confidence, given the great data limitations, yet be sure if the inter-industry linkages of Calgary manufacturers with Alberta industries are becoming stronger or weaker, more specific or more diverse. The effect of decline in exploration and development in the provincial conventional petroleum industry on Calgary's inter-industry manufacturing linkages can only approximately be surmised. Similarly the effect of change in other industrial regions of the world on Calgary's (or Alberta's) spatial industrial linkage cannot be clearly identified. These processes, and many others, must, however, be identified and considered before many of the problems facing the Calgary and Albertan economies can be properly identified and effectively solved.

Implications for Policy

The relatively weak inter-industry connections of Calgary and Albertan manufacturers suggest that many of the loosely stated proposals for industrial decentralization in the province and for discouragement of industry away from the major urban areas of Calgary and Edmonton do not take cognizance of the relative fragility of the manufacturing base in these two urban places. Where provincial inter-industry linkages with

Calgary manufacturers are strong--as in the case of the food and beverage industry, and the non-metallic mineral industry--decentralization proposals probably could be effected without excessive additions to the production costs of either Albertan or Calgary processors. But in those industries where extra-regional linkages have some crucial or overwhelming importance (non-food and general manufacturing, wood, paper and allied industries, primary metals, engineering, and transportation equipment) the added cost of assembly and distribution caused by location in places which might not have satisfactory access to trans-continental transportation services, warehousing or brokerage facilities, etc. could cause unreasonable hardship to industries which now find location in Calgary satisfactory or advantageous. And what of those numerous manufacturing industries which are not now found even in the larger Alberta places like Calgary? If economic forces operating in the Canadian or provincial economy have kept them away from Alberta then administrative and political plans for intra-provincial decentralization outside of the two largest Albertan places must be very carefully formulated indeed.

The linkages described in this paper reflect a multitude of economic and political forces which have been operating for a considerable time in the Canadian and world economies. They also reflect the demands of a sparsely populated region such as Alberta for a host of industrial commodities which characterize industrial societies at this point in the twentieth century. However, the highly generalised patterns described in this paper probably mask a great many diverse, and possibly contradictory, processes at work within each industry and within each firm in that industry. Since the underlying forces at work determining these linkage patterns are not clear or unambivalent, and since the patterns themselves are highly complex although the aggregate manufacturing economy may be relatively unsophisticated, any measures undertaken to intensify the Calgary manufacturing economy or to alter the spatial location of components of that economy must carefully incorporate some of the specific locational characteristics and spatial linkages associated with the industries comprising the groups analyzed in this paper.

CONCLUSIONS

The present study is but one component of a larger work investigating the nature of manufacturing linkages within the Alberta spatial economy, and the perception of locational opportunities and realities by Alberta industrial decision makers. The present study, however, raises far more questions than it answers. The study is based on data for which there are no published equivalents. The data themselves, however, are suggestive rather than definitive because of the variation in the way in which questionnaires were understood and completed by the respondents. Some of the specific observations in this paper may in fact reflect a particular data base rather than the objective structure of Calgary's manufacturing . However, notwithstanding some of these limitations, and despite the relative simplicity of the questionnaire, the nature of Calgary's backward and forward inter-industry provincial linkages has been described and tentatively evaluated. But there are many avenues of enquiry which could, and should, be followed from here.

Knowledge of the purpose and nature of Calgary's inter-industry linkages could be improved through investigation of the influence of non-Calgary ownership and control on city firms, of the role of external economies here and in other industrial regions, of the influence of friction of distance on the location of secondary industry in Alberta, and of the size of the local and provincial market for specific industrial products of Calgary firms.

Many studies have sought to explain industrial linkage in specific industries in which locational processes, for example, could be identified and reasonably explored. But attempts to see the entire urban-manufacturing pattern as an aggregate must, by necessity, group and amalgamate individual industries, losing sight thereby of some of the important underlying processes which may operate in some industries and not in others. Studies of aggregate urban linkages also suffer from an absence of theory or body of working hypotheses from which certain events may be expected or predicted with reasonable certainty.

Studies of aggregate linkages for individual urban places such as Calgary are in fact concerned with unique situations although the ultimate goal is identification and explanation of general processes and patterns. The present study has but scratched the surface of the problems associated with the spatial inter-connectivity of Calgary's manufacturing industries. Emphasis has been placed on the intra-regional, that is on intra-provincial linkages although the subject begs consideration of national and international linkages as well. Nothing has been said about intra-urban manufacturing linkages in Calgary although this subject would logically have to be investigated if attention were focussed on the problem of agglomeration economies within Calgary.

Although many of the subjects requiring further study will have to await the formation of a more comprehensive data base, data collected as part of this study will be used in forthcoming works to analyze the nature of regional inter-industry linkages for all of Alberta, and to assess the way in which different sectors and problems in the Alberta spatial economy are perceived by those entrepreneurs whose firms provided data for the linkages analyzed in the present paper.

APPENDIX 1

COMPOSITION OF INDUSTRIAL GROUPS IN THIS STUDY*

GROUP NUMBER	GROUP DESCRIPTION	S.I.C. GROUP CODE

(Component Industries Described According to the Standard Industrial Classification)

Group 1: Food and Beverage Industries.

Meat Products Industries	101,103
Dairy Products Industries	105,107
Fruit and Vegetable Canners and Preservers	112
Grain Mills	123,124,125
Bakery Products Industries	128, 129
Other Food Processors	131,133,135, 139
Beverage Manufacturers	141,143,145
Agricultural Implement Industry	311
Artifical Ice Manufacturers	3998

Group 2: Petrochemical, Coal, and Synthetic Textile Industries.

Rubber Industries	163,169
Synthetic Textile Mills	201
Petroleum and Coal Products Industries	365,369,371, 373,375,376, 378,379

Group 3: Non-Food, and General Manufacturing Industries.

Leather Industries	172,174,175, 179
Woolen Mills	193
Textile Industries (excluding synthetic textile mills)	213,221,223, 229
Knitting Mills	239
Clothing Industries	243,246,247
Furniture and Fixture Industries	261,264,266
Printing, Publishing and Allied Industries	286,287,288, 289

Machinery Industries (excluding agric. implement industry)	315,316
Electrical Products Industries	335,336,337, 338,339
Miscellaneous Manufacturing Industries- including the following:	
Scientific and Professional Equipment Manufacturing;	381
Jewellery and Silverware Manufacturing;	382
Broom, Brush and Mop Industry;	383
Venetian Blind Manufacturing	384
General Plastic Fabricators;	385
Sporting Goods and Toy Industry	3931
Fur Dressing and Dyeing Industry;	395
Signs and Displays Industry;	397
Model and Pattern Manufacturing;	3985
Fountain Pen and Pencil Manufacturing;	3989
Stamp and Stencil (Rubber and Meat) Manufacturers;	3995
Statuary, Regalia and Novelty Manufactureres;	3996

Group 4: Wood, Paper, and Allied Industries.

Wood Industries	251,252,254, 256,258,259
Paper and Allied Industries	271,272,273, 274

Group 5: Primary Metal, Engineering, and Transportation Industries.

Iron and Steel Mills	291
Pipe and Tube Mills	292
Iron Foundries	294
Non-ferrous Metal Smelters, Refineries, Rollers, Casters, Extruders	295,296,297, 298
Metal Fabricating Industries (excluding machinery industries)	301,309
Transportation Equipment Industries	321,324,325, 328,329

Group 6: Non-Metallic Mineral Products Industries

Manufacturers of:

Cement	341

Lime	343
Gypsum Products	345
Concrete Products	347
Ready-mix Concrete	348
Clay Products	351
Stone Products	353
Mineral Wool	354
Glass	356

*Note: Industries listed in the S.I.C. Code which have not established in Alberta are omitted from this Table.

APPENDIX II

STUDY OF LINKAGES IN THE MANUFACTURING
ECONOMY OF ALBERTA

QUESTIONNAIRE
(PART I)

1. Total Value of Manufactured or Semi-manufactured Materials and Supplies Used by the Plant: (See footnote 1)

____

2. Per cent of Total Materials and Supplies Used by the Plant but Manufactured in Alberta: (See footnote 2)

____

3. Total Selling Value of Factory Shipments from the Plant:

____

4. Per cent of Total Selling Value of Factory Shipments destined to all Customers within Alberta: (See footnote 2)

(This includes manufacturing and non-manufacturing customers).

____

Footnote 1:

This includes manufactured or semi-manufactured goods obtained directly from the manufacturer or through intermediary suppliers (e.g. wholesalers), such as sheet metal, scrap metal, basic chemicals, lumber, paper and paper products, acids, and all capital goods including equipment. This excludes purchases of raw or unprocessed materials (e.g. crude oil, natural gas, roundwood, grain, livestock, gypsum, and phosphate rock).

Footnote 2:

Please indicate approximate percentages even if actual amounts are unknown.

(PART 2)

MANUFACTURING GROUP:

MOVEMENT	(1) Food & Beverage	(2) Petrochemical Coal & Synthetic Textiles	(3) Non-Food, & General Manufacturing	(4) Wood, Paper Allied Industries	(5) Primary Metal, Engineering & Transportation	(6) Non-metallic Mineral Products
Of the materials and supplies originating within Alberta, what % came from each of these groups..						
Of the factory shipments destined within Alberta, what % went to manufacturers in each of these groups........						

REFERENCES

1. This study was financed by a grant from the Canada Council, and from The University of Calgary Grants Committee. The author gratefully acknowledges the assistance of Mrs. Eileen Shoesmith, Miss Janet March, Mr. Patrick Cadden, and Dr. Kenneth J. Fairbairn during the long period in which this research was undertaken. To all those members of the business community who gave of their time in completing the questionnaire, the author is especially indebted.

2. BRITTON, J.N. "A Geographical Approach to the Examination of Industrial Linkages," The Canadian Geographer, 13, No. 3 (1969), pp. 185-198; GILMOUR, J.M. "Some Considerations of Spatial Separation between Linked Industries," The Canadian Geographer, 15, No. 4 (1971), pp. 287-294; KARASKA, G.J. "Manufacturing Linkages in the Philadelphia Economy: Some Evidence of External Agglomeration Forces," pp. 256-267 in BOURNE, L.S. (ed.) Internal Structure of the City. New York: Oxford University Press, 1971.

3. HODGE, G. and WONG, C.C "Adapting Industrial Complex Analysis to the Realities of Regional Data," Papers, The Regional Science Association, 28 (1972), pp. 145-166.

4. STEED, G.P.F. "Changing Linkages and Internal Multiplier of an Industrial Complex," The Canadian Geographer, 14, No. 3 (1970), pp. 229-242; "The Northern Ireland Linen Complex, 1950-1970," Annals, AAG, 64, No.3 (1974), pp. 397-408.

5. BRITTON, op. cit.

6. KARASKA, op. cit.

7. RICHTER, C.E. "The Impact of Industrial Linkages on Geographic Association," Journal of Regional Science, 9, No. 1 (1969), pp. 19-28.

8. STREIT, M.E. "Spatial Associations and Economic Linkages Between Industries," Journal of Regional Science, 9, No.2 (1969), pp. 177-188.

9. GILMOUR, op. cit.

10. CZAMANSKI, S. Study of Clustering of Industries. Halifax: Institute of Public Affairs, Dalhousie University 1974.

11. SCHMIDT, C.G. "Firm Linkage Structure and Structural Change: A Graph Theoretical Analysis," Economic Geography, 51, No. 1 (1975), pp. 27-36.

12. THOMPSON, W. A Preface to Urban Economics. Baltimore: The Johns Hopkins Press, 1965; "Internal and External Factors in the Development of Urban Economies," pp. 43-62 in PERLOFF, H.S. and WINGO JR., L. (eds.) Issues in Urban Economics. Baltimore: Johns Hopkins Press, 1968.

13. BEYERS, W.B. "On Geographical Properties of Growth Centre Linkage Systems," Economic Geography, 50, No. 3 (1974), pp. 203-218.

14. CAMPBELL, J. "Selected Aspects of the Inter-industry Structure of the State of Washington, 1967," Economic Geography, 50, No. 1 (1974), pp. 35-46.

15. MOSELEY, N.J. "The Impact of Growth Centres in Rural Regions-I," Regional Studies, 7, (1973), pp. 57-75.

16. LEVER, W.F. "Industrial Movement, Spatial Association and Functional Linkages," Regional Studies, 6, (1973), pp. 387-400.

17. HODGE and WONG, op. cit.

18. STEED, op. cit.

19. SCHMIDT, op. cit.

20. TAYLOR, M.J. "Local Linkage, External Economies and the Ironfoundry Industry of the West Midlands and East Lancashire Conurbations," Regional Studies, 7, (1973), pp. 387-400.

21. SCHMIDT, op. cit., p. 27.

22. THOMPSON, op. cit., 1968, p. 45.

23. Ibid, p. 46.

24. PERLOFF, H.S. et al. Regions, Resources and Economic Growth. Lincoln: University of Nebraska Press, 1960, p. 395; Britton, op. cit. , p. 186.

25. Task Force on Urbanization, Index of Urban and Regional Studies Province of Alberta, 1950-1974. Edmonton, 1975.

26. WRIGHT, R.W. "The Alberta Economy - An Input/Output Analaysis." Report prepared for Calgary Power. 1966, 64 pages.

27. Ibid., Figure 7.

28. HANSON, E.J. Dynamic Decade (The Evolution and Effects of the Oil Industry in Alberta). Toronto: McClelland and Stewart, 1958, "Regional Employment and Income Effects of the Petroleum Industry in Alberta," in Proceedings of the Council of Economics, American Institute of Mining, Metallurgical and Petroleum Engineers, New York, 1966, 38 pages.

29. DRUGGE, S.E. "The Location of the Energy Intensive Industries," Alberta Human Resources Research Council, 1972.

30. SMITH, P.J. "Calgary: A Study in Urban Pattern," Economic Geography 39, No. 4 (1962), 15 pages; "Fort Saskatchewan: An Industrial Satellite of Edmonton," Plan, 3, No. 1 (1962), 13 pages

31. LEIGH, R. and CARTER, d. "Spatial Pattern of Economic Growth in Alberta, 1961-1966," The Albertan Geographer, No. 8, (1972), 7 pages.

32. SEIFRIED, N.R., "An Analysis of Recent Changes in Manufacturing in Alberta," The Albertan Geographer, No. 5 (1969), 6 pages.

33. BARR, B.M. "Has the Prairie Region Solved its Economic Problems?," British Columbia Geographical Series, 16, (1972), pp. 111-121; "Post-War Economic Reorganization of the Canadian Prairie Economy," pp. 65-82 in P.J. SMITH (ed.) The Prairie Provinces. Toronto: University of Toronto Press, 1972.

34. ZIEBER, G.H. "Inter- and Intra-City Location Patterns of Oil Offices for Calgary and Edmonton, 1950-1970," Unpublished Doctoral Dissertation, University of Alberta, 1971.

35. CURTIS, P.J. "Some Aspects of Industrial Linkages in Edmonton's Oil Industry: With Special Reference to the Tertiary Sector," Unpublished Master's Thesis, University of Alberta, 1972.

36. WILSON, L.S. "Some Factors Relating to the Attraction of Manufacturing Industries to the Province of Alberta," Unpublished Master's Thesis, University of Alberta, 1971.

37. BARR, B.M. and FAIRBAIRN, K.J. "Resource Extraction and the Generation of Manufacturing Linkages among Albertan Regions," in preparation.

38. WRIGHT, op. cit.

PLATE 7
Manufacturing of petroleum-related products, southeastern Calgary.

Shaw Pipe Protection, Ltd. Photo

CHAPTER 2

SOME ASPECTS OF CALGARY'S ROLE IN THE INTRA-INDUSTRY MANUFACTURING LINKAGES OF SOUTHERN ALBERTA: THE CASE OF THE MOBILE-HOME INDUSTRY

Alice McEwen
and
Brenton M. Barr

University of Calgary

INTRODUCTION

Objectives

The existence of two major urban places in Alberta suggests that the provincial urban-economic system is divided into two major spheres of influence. Excluding province-wide functions performed in Edmonton by the civil service for general government administration and in Calgary by head offices for the petroleum industry, the similarity of economic functions performed in both cities suggests that the province consists of two functional subregions with Edmonton dominant in the north and Calgary dominant in the south. Calgary's strong role in southern Alberta, as illustrated in Table 1,2 is attested to by the number of such important economic activities as wholesaling, and aggregated manufacturing production. The large share of these regional phenomena possessed by Calgary clearly demonstrates a statistical dominance of the city over all other places in southern Alberta and suggests that Calgary has a functional dominance as well.

On the basis of empirical urban research studies and general notions of central place theory that utilize higher order functions to ascertain tertiary industrial dominance within an urban system, Calgary may be described as dominating tertiary industry in southern Alberta. However, does this functional dominance of tertiary industry also pertain

TABLE 1,2

SELECTED ASPECTS OF THE ECONOMIC FUNCTION OF THOSE URBAN PLACES IN SOUTHERN ALBERTA ASSOCIATED WITH THE MOBILE-HOME INDUSTRY

Place	Total number of firms	Number of firms engaged in		Population
		manufacturing	wholesale trade	
Airdrie	30	2	3	1,090
Bowden	22	1	2	560
Brooks	135	10*	11	3,990
Calgary	5,468	511*	878	403,320
Cardston	83	4	4	2,685
Claresholm	90	12*	6	2,935
Coalhurst	2	1	-	426
Didsbury	68	4*	3	1,820
Fort MacLeod	86	6	6	2,715
Lethbridge	825	80*	93	41,215
Linden	22	4*	2	230
Medicine Hat	491	49*	32	26,520
Red Deer	544	37*	61	27,675
Standoff	2	1	-	4,785**
Stavely	13	1	2	350
Strathmore	45	2	2	1,150
Three Hills	59	4*	2	1,355
Total	7,985	729	1,107	522,821
% of Calgary in Total	68	70	79	77

Sources: Population, 1971 Census of Canada: "Population," Special Bulletins 92-772(SP-2), 1973, and 92-771(SP-1), 1973.

* Data from Alberta, Industry and Resources, 1973 (Edmonton: The Alberta Bureau of Statistics, 1973), pp. 20-21; all other data pertaining to number of firms are from Dun and Bradstreet Reference Book, 1971 (Toronto: 1971). "Total number of firms," in each place has been compiled from both these sources of information.

** Figure represents population on the Blood Indian Reserve.

to secondary industry, that is, to manufacturing? The large number of functions found in Calgary (Table 2,2) is evidence that the city is far more industrialized and performs a greater number and variety of functions than any other urban place in southern Alberta. We may account for this aspect of Calgary's dominance by reference to historical events, to the spatial structure of the transportation system, and to agglomeration economies. To the extent that products from industries located in Calgary are marketed in the region to private consumers, to other industrial operations, or to the construction industry, then we may conclude that:

1. Calgary has a functional manufacturing dominance within the region.
2. Certain economic advantages to manufacturers exist in Calgary but not elsewhere in the region.

The magnitude and structure of manufacturing in Calgary relative to those of other regional places are often cited as evidence that Calgary dominates the industrial structure of southern Alberta.

Industrial structure, however, conveys the meaning of functional linkages among manufacturers. Calgary's real dominance in the secondary industrial structure of southern Alberta thus should be confirmed by evidence of Calgary's commanding position and role in regional inter-industry linkages or in regional intra-industry manufacturing linkages. Since problems associated with obtaining data are so severe that <u>inter-industry</u> linkages can be determined adequately only between Calgary and the Province of Alberta as a whole (see Barr, Chapter 1), this paper will attempt instead to examine Calgary's intra-industry manufacturing linkages within southern Alberta in order:

1. To describe, and provide tentative explanation of, Calgary's

TABLE 2,2

SELECTED STRUCTURAL COMPONENTS OF THOSE MAJOR URBAN PLACES IN SOUTHERN ALBERTA ASSOCIATED WITH THE MOBILE-HOME INDUSTRY, 1970

PLACE	MANUFACTURING FIRMS			EMPLOYMENT IN MANUFACTURING			VALUE ADDED BY MANUFACTURERS		
	Number	% of Region	% of Alberta	Number of Employees	% of Region	% of Alberta	$\$ \times 10^3$	% of Region	% of Alberta
Medicine Hat-Redcliff	49	5.9	2.5	1900	8.5	3.8	19.7	6.6	2.9
Census Division 1	49	5.9	2.5	1900	8.5	3.8	19.7	6.6	2.9
Lethbridge	80	9.6	4.1	2627	11.8	5.2	34.7	11.6	5.0
Brooks	10	1.2	.5	49	.2	.1	.5	.2	.1
Census Division 2	118	14.1	6.0	3325	14.9	6.6	44.1	14.7	6.4
Claresholm	12	1.4	.6	288	1.3	.6	1.9	.6	.3
Census Division 3	35	4.2	1.8	561	2.5	1.1	4.0	1.3	.6
Three Hills	4	.5	.2	11	---	---	.07	---	---
Linden	4	.5	.2	34	.2	.1	.3	---	---
Census Division 5	23	2.7	1.2	111	.5	.2	.8	.3	.1
Calgary	511	61.1	26.1	15246	68.5	30.3	212.2	70.8	30.7
Didsbury	4	.5	.2	12	.1	---	.07	---	---
Census Division 6	547	65.4	28.0	15472	69.5	30.7	214.9	71.7	31.1
Red Deer	37	4.4	1.9	726	3.3	1.4	13.5	4.5	1.9
Census Division 8	60	7.2	3.1	876	3.9	1.7	15.9	5.3	2.3
Regional Total (ie. "Southern Alberta")*	837	100.0	42.8	22265	100.0	44.2	299.6	100.0	43.4
Alberta Total	1956		100.0	50352		100.0	690.1		100.0

* Includes C. Division 4.

Source: Alberta Industry and Resources 1973. Department of Industry and Commerce, Alberta Bureau of Statistics. Edmonton, Alberta, pp. 20-21.

role in these linkages.

2. To determine if Calgary does, in fact, exert a regional influence in manufacturing which is proportionate to its influence in tertiary industry.

3. To identify the possible existence of a regional manufacturing complex in southern Alberta.

Conclusions from the three basic objectives may have significant implications for those engaged in determination of urban and industrial economic policies for the province. Before explicit attempts are made to divert growth from Calgary (or Edmonton), a better understanding of this city's regional manufacturing function and relationship with other regional manufacturers and urban places is needed. As yet, Calgary's influence in manufacturing is unknown because no clear statements exist as to whether Calgary is an independent manufacturing entity, or whether it has important interdependence with other producers and places in southern Alberta. Further, policies related to an industrial complex characterized by specialization and significant intra-industry absorption of manufactured goods should differ significantly from those associated with a nascent or tenuous regional manufacturing structure comprised chiefly of intra-industrial linkage with extra-regional manufacturers and industrial complexes. Therefore, careful analyses should be carried out by policy-makers to identify those industries, which because of their intra-industrial locational structure, can safely withstand attempts to move them to non-metropolitan Alberta locations, and to identify those industries which seem to have an overwhelming attraction to metropolitan locations in Alberta.

Related Studies

Analyses of such external economies as mobile or immobile economies, and localization, urbanization, or spatial juxtaposition economies has produced a theoretical, conceptual, and empirical body of knowledge of great value. Many studies have demonstrated that firms, not otherwise dependent on locations at resource-extractive sites, do make access to producers of subcomponents (backward linkages) or to industrial consumers of manufactured goods (forward linkages) a prime concern in their selection of plant locations. Many economic advantages thus exist for individual firms when they locate in some form of spatial proximity to related manufacturers.

Most spatial analyses of external economies have been concerned with forms of agglomeration created within urban areas or general urban regions. However, as Britton[1] has pointed out, considerably less emphasis has been placed on analyses of production linkages between urban regions. If the proportion of mobile economies to immobile economies is increasing as Robinson[2] suggests, and if production chains are indeed becoming functionally more complex and spatially longer, then indeed more emphasis should be placed on analyses of inter-urban manufacturing linkages in order to identify more precisely the meaning and function of industrial structures. Significant flows of manufactured commodities between spatially separated manufacturing firms also suggest that internal economies of scale may in fact be an important process which offsets in many economies the supposed advantages of agglomeration. Thus linkages among manufacturers caused by firms "tending to minimize the opportunity costs involved in assembling or dispatching intermediate goods"[3] may not in fact result in close promimity but may lead to linkages throughout an economy. Linkage economies in this way "can become so important that they are internal economies to the complex" of spatially-separated interdependent plants.[4]

Depending on numerous micro-economic considerations within individual firms and technological factors associated with particular forms of production, patterns of interdependency and linkage may occur at different spatial scales: urban, regional, macro-regional and national or international. Furthermore, the spatial extent of such manufacturing interdependencies may be governed by the geographical size of the economy, the distances separating urban areas, the prevailing level of transportation costs and availability of service, and the distribution and density of the population. National attitudes toward multi-national corporations and international specialization also influence the complexity and extent of linkage patterns. And as Steed[5] has shown, the existence of manufacturing linkages also depends on the general state of an industry, its temporal contribution to society, and its ability to withstand competition. Consequently as the nature of manufacturing changes (technoligical change and product obsolescence) the linkages among urban places improve or decline. The influence of all these factors and processes on Calgary's role in the intra-industry structure of southern Alberta will be discussed in subsequent sections of this chapter.

Other analyses of manufacturing linkages have commonly relied either on aggregate economic data or, in a smaller number of cases, on empirical data for particular firms or industries. Aggregate and individual analyses often have been conducted, however, without a strong theoretical base on which to interpret their findings. Guthrie, in his review of economies of scale and regional development, noted that "no significant attempt has been made to merge the analytical framework of this portion of micro theory," that is economies of scale, "with locational theory or to apply the limited empirical findings on size of firm to the problems of regional development."[6] Subsequent research by numerous authors, for example Isard and Schooler,[7] which has helped remedy this situation also has revealed the difficulty in identifying economic linkages within specific manufacturing complexes, the problems in applying aggregate techniques such as input-output, economic base, commodity flow analysis, and multiple regression,[8] and the patience and skill required to derive linkage structures among numerous firms within a national industrial complex.[9] The normal problems of applying analytical techniques are further compounded by the dearth of published data, variation in product definition and measurement criteria from country to country, and in the high cost and poor rates of return associated with derivation of private data from questionnaires.

Although considerable advancement has been made in the general understanding of inter-industry relations, recent analyses of the strength of linkages between related industries confirm that inter-industry analysis is still short on theory and inadequately served by input-output and multi-variate analysis. Thus, the fundamental problem still remains of adequate accounting for patterns of locational association and linkage, and for the impact of distance on the spatial separation of linked manufacturing activities. Problems caused by the absence of a powerful theoretical and conceptual base to this area of research and enquiry are further compounded by still inadequate amounts of data.

Data Problems

Problems associated with inter-industry linkage encountered elsewhere and in other studies are no less acute in the functional analysis of Calgary's manufacturing association with other urban places in southern Alberta. The general dearth of published data is aggravated by the problem of confidentiality in Canadian economic statistics, (see Gilmour,[10]) caused by the disporportionately large share of a few manufacturers in the economy of small urban places. In addition, Canadian statistical agencies still do not collect locational data pertaining to the functional relations among manufacturing enterprises, or between manufacturing enterprises and those of other sectors of the economy. Consequently, origin and destination of materials purchased and shipped by manufacturers are impossible to derive from either published sources or from unpublished government materials even when they can be consulted. Evidence to identify Calgary's role with other urban places in southern Alberta (or anywhere, for that matter) must be sought through data derived directly from manufacturers themselves. This approach, however, is expensive and time-consuming; and its utility is completly dependent not only on the willingness of the entrepreneur to assist, but in the quality of his own information. Perhaps one reason why government is so reluctant to obtain data pertaining to spatial manufacturing linkages is because many entrepreneurs themselves do not know the proportion of value of their total purchases or sales accounted for by particular origins and destinations.

Manufacturers' locational data pertaining to inter-industry relations are more difficult to obtain than those related to intra-industry linkages. This difference appears to be related to the importance of wholesalers as intermediaries between firms in different industries, and to the general amount of knowledge and information which entrepreneurs tend to possess about the structure of their own industry. Therefore, since the objectives of this study are so closely associated with identification of locational linkages, the authors decided to base their observations and generalizations on the intra-industrial role of Calgary within its region on data obtained from entrepreneurs in one major industry, the Mobile-Home Industry.

Data Base for the Mobile-Home Industry

The Mobile-Home Industry is defined in this study as that industry producing mobile homes, relocatable accommodation, pre-fabricated buildings, travel and utility trailers, as well as related exterior, structural, and interior component parts. This definition, by including producers of the final product as well as producers of semi-finished components which enter the final product, incorporates direct and indirect producers who constitute the Mobile-Home Industry.

Interviews with manufacturers in the Mobile-Home Industry located within the region provided the data base for analysis of the Mobile-Home Industry. The initial data were collected as part of a major analysis of the influence of abandoned military bases on the location of industry in southern Alberta. Hence, the same data were utilized and adapted to fit the present study of those products produced by manufacturers on the site of each former military base. These products were grouped into five categories on the basis of production similarities:

1. Finished mobile-home products (mobile-homes, travel trailers, pre-fabricated buildings).
2. Exterior and structural component parts.
3. Interior component parts.
4. Agricultural implements.
5. Miscellaneous products.

The first three categories reflect the presence of Mobile-Home Industry firms which heavily influences the manufacturing structure on the bases. Therefore, only the responses from firms in these categories were used in the present study.

All manufacturers in the region making products similar to those produced on the bases, but not located on abandoned military sites, were identified from the Alberta Trade Index, 1972. These manufacturers were placed in the appropriate categories and a ten percent sample was used to ensure sufficient response to carry out the study. Executives of these firms and of those associated with all secondary manufacturing in the bases were interviewed. Therefore, the data used, in this analysis provide complete coverage of the Mobile-Home Industry located in abandoned military bases and representative coverage of that segment of the industry located elsewhere in the region. Other data gathered as part of the major study reveal firms of the Mobile-Home Industry located on old base facilities perform the same functions as those located elsewhere in the region, thus, eliminating any bias which may have resulted from emphasis on the bases.

The data were obtained from face-to-face interviews between the prime author and either the general manager or the president of the corporations visited. The questionnaire was structured (see Appendix I) but included a final section with an open-ended format. This study, however, uses data only from the structured portion of the questionnaire. Responses to the first question provided information on the value of commodities bought from or sold to wholesalers in Calgary, and the third question sought to determine the percentage of the value of goods entering or leaving the plant which are transhipped through Calgary. Since these data were provided from the general fund of knowledge about the firm possessed by each respondent, they are probably accurate in a general way but are not amenable to precise interpretation. Values, for example, of twenty per cent might in fact represent approximately a quarter, or a sixth but not exactly one-fifth. On the other hand, better data simply are not available from any other source, private or public. Like all studies of the firm in which confidential data are elicited, the industrial geographer simply must rely on the goodwill and cooperation of the respondent. The respondent usually does not possess data in the form requested by academic, industrial geographers. The authors, therefore, assume that the majority of those interviewed possessed information of similar quality about the linkages of their firm with other enterprises, and that the general information provided is sufficiently accurate to permit reliable and at least, objective, assessment of the intra-industry manufacturing linkages between firms in southern Alberta and those in Calgary.

Intra-industry Linkages of the Mobile-Home Industry

Nearly four per cent of manufacturers in Alberta are engaged in the final production of mobile-homes, and nearly another four per cent are engaged in production of subcomponents. Hence, nearly eight per cent of Alberta's manufacturers are members of the Mobile-Home Industry. The significance of this share is further enhanced when we realize that the largest number of manufacturers in this province (sawmills, bakeries, machine shops, dairies, food processors, printers, publishers, book-binders, and dental labs) are engaged in the production of uncomplicated and simple final products in relatively small establishments. Material obtained as part of a major study of inter-industry linkages in Alberta[11] confirms that these manufacturers (except for food processors and machine shops) have few linkages with manufacturing enterprises located in Alberta. The Mobile-Home Industry, on the other hand, has important manufacturing links and production chains within itself, and although many of these chains extend beyond the confines of this province due to the generally small market here for subcomponent products, the industry nevertheless does have some intra-Albertan linkages which distinguish it from many other manufacturers currently producing in the region (Figure 1,2).

The Location and Function of the Mobile-Home Industry in Southern Alberta

Calgary is the dominant center of the Mobile-Home Industry in southern Alberta (Table 3,2); the city dominates both in the number of establishments producing the final product (49 per cent) and in the number of subcomponent manufacturers (61 per cent). The next largest producer, Lethbridge, has approximately 13 per cent of the final producers and almost the same share of subcomponent producers. The share

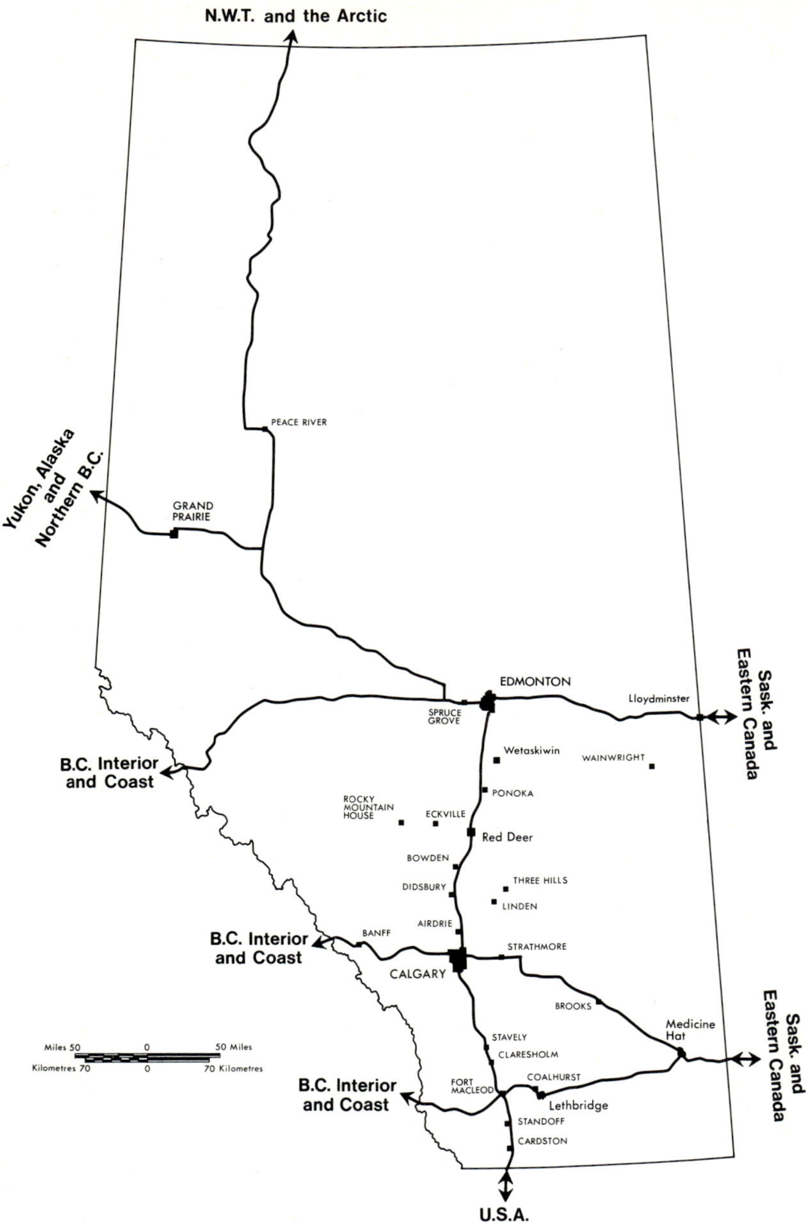

FIGURE 1,2 Location of Firms in the Mobile Home Industry.

TABLE 3,2

THE MOBILE-HOME INDUSTRY IN SOUTHERN ALBERTA

PLACE	NUMBER OF FIRMS			NUMBER OF FIRMS AS % OF TOTAL NUMBER OF MOBILE-HOME INDUSTRY FIRMS IN:					
				Southern Alberta			Alberta		
	Category I	Category II & III	Total	Category I	Category II & III	Total	Category I	Category II & III	Total
Medicine Hat	2	--	2	4.5	--	2.4	2.8	--	1.4
Census Division 1	2	--	2	4.5	--	2.4	2.8	--	1.4
Brooks	--	1	1	--	2.6	1.2	--	1.4	.7
Coalhurst	--	1	1	--	2.6	1.2		1.4	.7
Lethbridge	6	5	11	13.3	13.2	13.2	8.4	6.7	7.5
Census Division 2	6	7	13	13.3	18.4	15.6	8.4	9.5	8.9
Cardston	1	--	1	2.2	--	1.2	1.4	--	.7
Claresholm	1	3	4	2.2	7.9	4.8	1.4	4.0	2.8
Fort Macleod	2	--	2	4.5	--	2.4	2.8	--	1.4
Standoff	1	--	1	2.2	--	1.2	1.4	--	.7
Stavely	1	--	1	2.2	--	1.2	1.4	--	.7
Census Division 3	6	3	9	13.3	7.9	10.8	8.4	4.0	6.3
Linden	1	1	2	2.2	2.6	2.4	1.4	1.4	1.4
Strathmore	1	--	1	2.2	--	1.2	1.4	--	.7
Three Hills	1	--	1	2.2	--	1.2	1.4	--	.7
Census Division 5	3	1	4	6.6	2.6	4.8	4.2	1.4	2.8
Airdrie	2	--	2	4.5	--	2.4	2.8	--	1.4
Calgary	22	23	45	49.0	60.6	54.4	30.5	31.0	30.6
Didsbury	--	1	1	--	2.6	1.2	--	1.4	.7
Census Division 6	24	24	48	53.5	63.2	58.0	33.3	32.4	32.7
Bowden	1	--	1	2.2	--	1.2	1.4	--	.7
Red Deer	3	3	6	6.6	7.9	7.2	4.2	4.0	4.0
Census Division 8	4	3	7	8.8	7.9	8.4	5.6	4.0	4.7
TOTAL: SOUTHERN ALBERTA	45	38	83	100.0	100.0	100.0	62.7	51.3	56.8
TOTAL: ALBERTA	72	74	146				100.0	100.0	100.0

Source: Alberta Trade Index 1972. Alberta Bureau of Statistics, Edmonton, Alberta, 1972

of other southern Alberta places is even smaller. Calgary has approximately 54 per cent of the aggregate Mobile-Home Industry but Lethbridge, the second largest populated place in the region, has only 13 per cent. The remainder of the Mobile-Home Industry is spread among sixteen other places although the relative share of these places in the production of the final product is much greater than their share of subcomponent manufacturing (Table 3,2). Since Calgary dominates the locational pattern of manufacture of the final product and subcomponents within the Mobile-Home Industry, the question must be asked, does Calgary also dominate the pattern of functional linkages within the Mobile-Home Industry?

CALGARY'S ROLE IN THE MOBILE-HOME INDUSTRY

Intra-Mobile-Home Industry Linkages Between Calgary and Other Regional Places

Should patterns of market-orientation and local self-sufficiency be expected for this industry? The presence of a relatively larger share of subcomponent producers than final producers in Calgary suggests that Calgary might dominate the region in terms of supplying subcomponent or semi-manufactured products to other final producers in the region. The response, from manufacturers interviewed during field research suggests that, in fact, Calgary does not have strong manufacturing linkages with regional producers of mobile homes. There are, however, variations in the pattern which suggest that Calgary's relationship with other mobile home producers depends to a limited extent at least on the function of these producers within the general industry, and on the extent to which regional producers are small independent firms or members of national or international corporations.

The majority of firms in category one (final producers of mobile-homes) have some linkage with Calgary subcomponent manufacturers but the extent of this linkage is modest indeed (Table 4,2). This observation tends to confirm the notion that most regional places producing the final mobile-home product are relatively independent of Calgary subcomponent manufacturers and either rely on local producers or purchase the majority of their requirements from outside the region. In Claresholm, for example, Safeway Shelter Systems is located adjacent to subcomponent producers (Berdick Manufacturing, Trus-Joist (Western), and Auroral Tool) from whom it makes important purchases.

Subcomponent producers in regional places outside Calgary have linkages with Calgary manufacturers which are weaker than those of producers of the final product (Table 4,2). All six subcomponent producers failed to report any significant purchases from any manufacturers located in Calgary. We must tentatively conclude, therefore, that Calgary neither dominates intra-industry linkages within the Mobile-Home Industry in southern Alberta, nor generates significant manufacturing ties between any industry and regional manufacturers in the Mobile-Home Industry. Further confirmation of this conclusion for the province as a whole is found in examination of firms in Wetaskiwin and Edmonton (Table 4,2), places not considered part of the southern Alberta region. The responses of these firms also illustrate their lack of contact with Calgary manufacturers.

Should regional producers of mobile homes be expected to have strong linkages with consumers in Calgary given the overwhelming potential size of that market compared to that of the rest of the region? The response of the manufacturers shows that only Calgary producers have an extensive market in this city. All other producers, both of final products and component parts, sell less than twenty per cent of their product to manufacturers or consumers in Calgary. Calgary thus dominates neither the regional intra-industry absorption of mobile home pro-

TABLE 4,2

INTRA-INDUSTRY MANUFACTURING LINKAGES OF THE SOUTHERN ALBERTA MOBILE-HOME INDUSTRY WITH FIRMS LOCATED IN CALGARY

CATEGORY OF Mobile-Home Producer ↓	LOCALITY	PERCENT OF MANUFACTURED INPUT:			PERCENT OF MANUFACTURED OUTPUT:		
		Purchased From Calgary:		Transhipped Through Calgary	Sold To Calgary:		Transhipped Through Calgary
		Manufacturers	Wholesalers		Manufacturers Or Final Consumers	Wholesalers	
I:	Claresholm	10	5	50	--	--	--
	Claresholm	20	10	5	--	--	--
PRODUCERS	Fort MacLeod	10	2	--	--	--	--
of the	Fort MacLeod	--	--	--	--	--	15
FINAL	Fort MacLeod	30	16	45	--	75	--
PRODUCT	Lethbridge	10	15	5	--	--	--
	Lethbridge	--	10	10	--	10	--
	Linden	5	5	70	--	70	20
	Medicine Hat	5	20	70	--	20	--
	Red Deer	10	5	80	5	--	--
	(Extra-Regional Localities:						
	Edmonton	--	8	8	--	--	--
	Wetaskiwin)	15	4	--	--	20	--
II & III:	Calgary	5	80	--	--	NA	NA
	Calgary	--	75	80	--	NA	NA
PRODUCERS	Claresholm	--	20	--	--	--	--
of	Claresholm	--	--	--	--	--	--
COMPONENT	Fort MacLeod	--	--	--	--	--	--
and	Red Deer	--	--	--	--	--	--
STRUCTURAL PARTS	(Extra-Regional Localities:						
	Edmonton	5	--	2	--	--	--
	Edmonton	--	--	--	--	--	--
	Edmonton	--	3	1	--	--	--

Source: Questionnaire Survey, May - September, 1974
NA = Not Applicable

ducts and subcomponents, nor the demand by final consumers for the products of the southern Alberta Mobile-Home Industry.

Wholesale Linkages Within the Mobile-Home Industry Between Calgary and Other Regional Places

The numerical dominance of Calgary wholesalers in southern Alberta is devoid of meaning unless secondary industries within the region make use of their services. The share of these wholesalers in the total purchase of manufactured inputs by firms of the Mobile-Home Industry ranges from nil to eighty per cent. The component producers have no real contact with wholesalers as only one of the six firms makes use of their services; this firm is located in Calgary and purchases eighty per cent of its manufactured inputs from Calgary wholesalers. However, all final-product producers have some contact with wholesalers; these products comprise from five to eighty per cent of their purchases of manufactured commodities but the range varies so much among producers in each locality that no discernable pattern can be established. Thus, wholesalers appear to serve a greater function for final producers than for component producers; but the amount of contact appears to be determined by variables other than those measured in this study.

Responses by Westaskiwin and Edmonton firms also indicate an absence of significant contact with Calgary wholesalers. Edmonton, as a larger centre has more wholesalers than Calgary, so that manufacturers in the city who make use of wholesale services use these firms instead of those in Calgary.

Use of wholesalers for distribution of the final product in the Mobile-Home Industry is non-existent. All manufacturers of the final-product ship directly to retail outlets; component producers sell directly to other manufacturers on a contract basis. The overall pattern of wholesaling in the Mobile-Home Industry suggests that Calgary's wholesalers are oriented to retail and service industries rather than to manufacturing.

Mobile-Home Industry Reliance on Calgary as a Transhipment Centre

Calgary's locational situation suggests that the city should be a convenient transhipment point to the smaller urban centres in the province: Calgary has the greatest number of trucking firms in the region; it is located at the junction of the Trans-Canada Highway (east-west links) and Highway 2, Alberta's major north-south route; and it is served by both national railways and by international, national, and regional airlines. But do these factors significantly influence the movement of goods? Must trucking companies, for example, tranship at Calgary if the destination of their major cargoes is to a regional centre located on the highway network? Furthermore, to what extent is each of the transport media used by any industry?

Response from the firms questioned shows that, with two exceptions, Calgary is not used as a major transhipment point; shipments from extra-regional suppliers are usually made directly to manufacturers in smaller centres. One exception is Linden, located off the major provincial highway system; the second exception involves unloading of goods shipped to Alberta by rail in less-than carload lots, a form of transport seldom used by the Mobile-Home Industry.[12]

The nature of the final product dictates that direct shipping by truck is the common form of transport within the Mobile-Home Industry. Calgary is not used as a transhipment centre for this purpose either. In most cases, the market area consists of the three westernmost provinces, and distances to market do not warrant extensive use of rail transport by firms in the Mobile-Home Industry. Further, since most national and region trucking firms serve the major communities throughout the region on the basis of an area-wide freight-rate structure, there is no need to

PLATE 8
Light manufacturing chiefly serves the local and regional market.

Calgary Tourist and Convention Assn. Photo

PLATE 9
Eastern Calgary's industries are associated with petroleum and food processing.

City of Calgary Photo

tranship through Calgary

DISCUSSION

The Mobile-Home Industry grew in Alberta as a result of oil exploration and construction; mobile and prefabricated units are now used most extensively in the northern part of the province while Calgary serves as a centre of petroleum head offices. Thus, Calgary firms and consumers don't serve as significant producers or consumers of processed goods in the regional Mobile-Home Industry. Mobile-Home Industry products are shipped to exploration and development sites (eg. northern Alberta, the Mackenzie Valley, the Arctic archipelago) rather than to Calgary itself. In one segment of the industry, recreational vehicles, Calgary possesses a small potential market although its influence on the industry does reflect its proximity to recreational areas in the Rocky Mountains.

However, as a wholesaling centre, Calgary exerts a greater influence than as a manufacturing centre although wholesaling is not so directed toward secondary industry as toward the tertiary sector. This aspect of Calgary's economic activity is a consequence of the sparse population throughout the entire region. The small population, the economic importance of petroleum extraction and export, and the extensive development of agriculture in Alberta cause urban places for the province in general, and Calgary in particular, to function as service centres rather than as manufacturing complexes.

Calgary's spatial location, and its size and diversity in transportation services, do not appear to have combined to give the city a role as a major transhipment point for regional manufacturers. Although this may only reflect the nature of the Mobile-Home Industry as a primary user of truck rather than rail or air services, the basic structure of manufacturing in southern Alberta appears to preclude the need for Calgary to serve as a transhipment centre for regional manufacturers.

Given the general character of sparsely populated, resource-oriented economies, urban areas within the region serve primarily as service centres; their limited manufacturing functions are dependent on producers in complex industrial regions. Evidence obtained through this study of Calgary's role in intra-industry linkages supports this general notion and appears to have wider applicability than just to the Mobile-Home Industry. This study points to a lack of intra-industry manufacturing linkage between Calgary and its region, and to the general absence of a regional manufacturing complex in southern Alberta. The influence exerted by Calgary over secondary industry in its region is, therefore, minimal.

Furthermore, in the larger study by McEwen of abandoned military establishments, industrial complexes were identified on each of the bases. The manufacturing industries in each of the urban centres of southern Alberta have very little regional inter-firm linkage beyond their own complexes. These industries display, however, excessive dependence for most of their manufactured component parts on the large and complex industrial regions of eastern Canada and the United States. Calgary's role in regional intra-industry and inter-firm manufacturing linkages may, therefore, be considered as incidental to the city's general economic function.

Within the broad perspective of Calgary's role in the structure of manufacturing in southern Alberta, this study suggests that, although Calgary is a large urban centre in comparison to other places in its region, it has little impact on regional secondary industry. Calgary in many ways seems to be simply a larger example of the general type of urban-manufacturing economies found throughout the region although the size of population and labour force allows for greater variety in Calgary's manufacturing industries than in those of other regional places; each sector of the Calgary manufacturing economy appears from this study to be relatively independent of similar sectors located in other southern Alberta urban places.

CONCLUSIONS

The general industrial function of Calgary's secondary industry is to manufacture products in support of the major primary industries engaged in resource exploitation in Alberta. This role includes production of implements and equipment for agriculture; processing of farm produce; manufacture of prefabricated structures for the petroleum and construction industries; and production of numerous small, market-oriented products required by the population directly or indirectly dependent on these primary industries.

Within the manufacturing economy, intra-industry and inter-firm linkages suggest that Calgary is not a dominant centre, and its structure is relatively independent of, but parallel to, that of other urban places in southern Alberta.

No regional centre exerts an excessive manufacturing influence on any other regional centre; all depend on extra-regional industrial complexes. Thus, regional policies for industrial and economic development should become more cognizant that an inter-urban, intra-industry, manufacturing complex does not exist in southern Alberta.

APPENDIX I

QUESTIONNAIRE

1. What percentage of your manufactured inputs is supplied by firms who manufacture the product in Calgary?

2. What percentage of your product is consumed in Calgary?

3. What percentage of your manufactured inputs is obtained from wholesalers in Calgary?

4. What percentage of your product is sold through wholesalers in Calgary?

5. What percentage of your manufactured inputs is transhipped to you through Calgary?

6. What percentage of your product is transhipped to your customers through Calgary?

REFERENCES

1. BRITTON, J.N. "A Geographical Approach to the Examination of Industrial Linkages," The Canadian Geographer, 13, No. 3 (1969), pp. 185-198.

2. ROBINSON, E.A.G. The Structure of Competitive Industry. Chicago: The University of Chicago Press, 1964, p. 124.

3. BRITTON, op. cit., p. 185.

4. Ibid.

5. STEED, G.P.F. "Changing Linkages and Internal Multiplier of an Industrial Complex," The Canadian Geographer, 14, No. 3 (1970), pp.229-242; "The Northern Ireland Linen Complex, 1950-1970," Annals, AAG, 64, No.3 (1974), pp. 397-408.

6. GUTHRIE, J.A. "Economies of Scale and Regional Development," Papers and Proceedings, RSA, 1 (1955), p. J-2.

7. ISARD, W. and SCHOOLER, E.W. "Industrial Complex Analysis, Agglomeration Economies, and Regional Development," Journal of Regional Science, 1, No.2 (1959), pp. 19-33.

8. BRITTON, op. cit.

9. STEED, op. cit.

10. GILMOUR, J.M. "The Joint Anarchy of 'Confidentiality' and Definitional Change," The Canadian Geographer, 10, No. 1 (1966), pp. 40-48.

11. BARR, B.M. and FAIRBAIRN, K.J. "Resource Extraction and the Generation of Manufacturing Linkages among Albertan Regions," in preparation.

12. MCEWEN, A. "Reuse of Deactivated Military Bases as Industrial Parks," Master's thesis in preparation, University of Calgary.

CHAPTER 3

CALGARY AS AN OIL ADMINISTRATIVE AND OIL OPERATIONS CENTRE

G.H. Zieber

University of Lethbridge

Calgary is Canada's outstanding oil administrative centre. Situated in the nation's primary oil and natural gas producing region (Figure 1,3), Calgary is the home of the largest number of its administrative and operations oil offices and in particular head offices. Development as an oil centre until 1947 was marked by slow growth, as well as sporadic boom and bust conditions, associated with three periods of discovery of oil and natural gas in the Turner Valley. It was the important Leduc oil discovery, in 1947, and subsequent major finds elsewhere which led to the remarkable growth of Calgary's oil function and transformed the city into one of Canada's two leading oil centres. The other large centre with an important but essentially operations oil function is its neighbour to the north, Edmonton.

CALGARY'S DEVELOPMENT AS AN OIL CENTRE

It was the discovery of oil and natural gas in 1914 by the Calgary Petroleum Products Company in the Turner Valley approximately twenty-five miles southwest of Calgary which gave birth to Calgary's function as an oil centre. Because Calgary was the only major urban community close to Turner Valley and could provide the services and facilities required, it experienced a great influx of oil personnel and equipment particularly from the United States and became a booming oil city. Hundreds of oil companies were formed, brokerage houses sprang up throughout the city, and a stock exchange was organized. However, this inital boom lasted but three months. The majority of the oil companies later exited from the business or became inactive.

The Calgary Petroleum Products Company which survived the collapse of the boom was taken over in 1920 by the Royalite Oil Company, a subsidiary of Imperial Oil. In 1923, Imperial Oil began operating Alberta's first modern oil refinery in Calgary with Montana and Wyoming being the chief source of crude.[1] Royalite's drilling of a successful well at Turner Valley in 1924, containing very light oil and wet gas, resulting in a renewed burst of drilling activity with more than two hundred wells drilled during the next twelve years. Again Calgary experienced boom conditions and derived the benefits of the oil activity. The number of oil companies (producers, explorers, developers) rose from 13 in 1920 to 23 in 1925, and 56 in 1930 (Table 1,3). By 1930 Calgary had a total of 198 various companies associated with the petroleum industry.

It was not until 1936, that the major oil pool of the Turner Valley field was penetrated. Three Calgary businessmen, R.A. Brown, Superintendent of the Calgary Electric Light and Power Department, G.M. Bell, a Calgary newspaper owner, and J.M. Moyer, pooled their life savings, and with contributions from Imperial Oil and the British American Oil Company, drilled the discovery well. Drilling activity again was great and by 1942 oil companies had brought in nearly 200 wells.[2] Table 1,3 shows the renewed rise in oil firms for Calgary by 1940.

In order to avoid overproduction and consequent waste of oil in the Turner Valley field, Imperial Oil, operator of the only Calgary refinery, set up a system of quotas (termed prorationing[3]) to be purchased from producers. The producers, many of them small independent com-

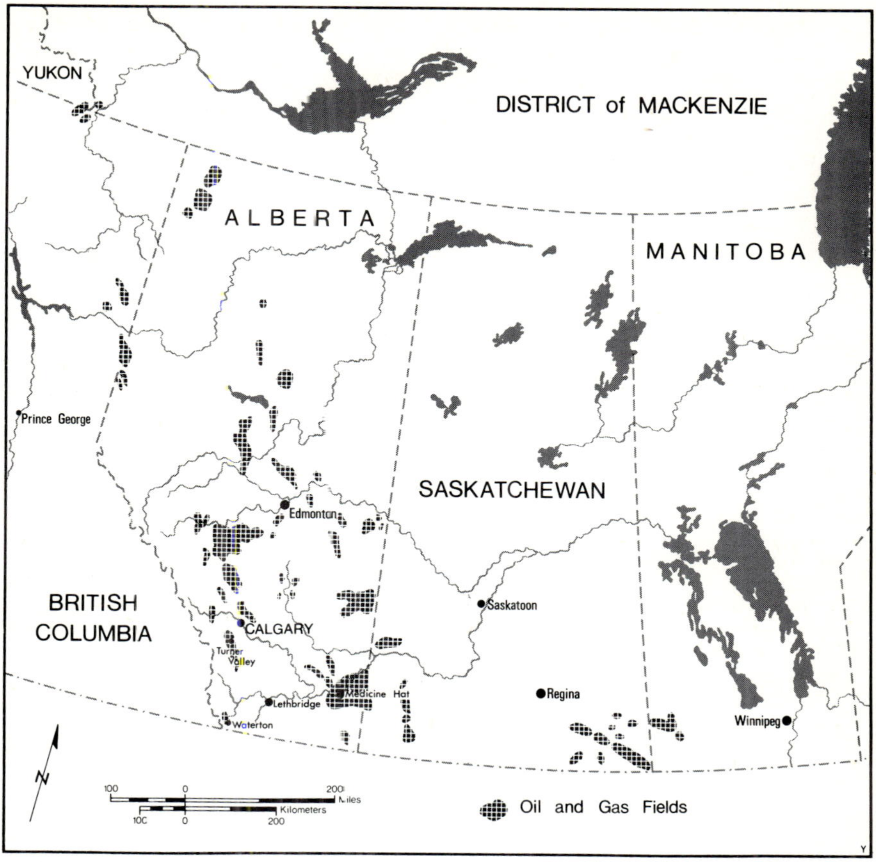

FIGURE 1,3 Situation of Calgary.

panies, objected to having a major firm restrict their production. To deal with the situation, the Alberta Government set up the Petroleum and Natural Gas Conservation Board in 1938 with headquarters in Calgary to allocate quotas and to regulate oil and gas production in the interests of conservation. By choosing Calgary as the headquarters for the Board, the government gave recognition to Calgary's role as the centre of the petroleum industry at that time. In 1939, Calgary

TABLE 1,3

OIL OFFICES IN CALGARY PRIOR TO 1947

Oil Office Types	1920	1925	1930	1935	1940	1945
Oil Companies (producers, explorers, developers)	13	23	56	37	56	62
Geologists, Geophysicists	-	-	4	2	3	4
Oil and Mining Stocks	7	-	31	19	19	11*
Oil Exchanges	-	-	2	1	-	-
Oil Brokers and Land Agents	-	-	75	15	21	11
Oil and Gas Well Supply, Equipment, and Service	-	-	13	6	15	17
Oilwell Drillers	1	-	8	1	5	14
Oil Pipe Lines	-	-	-	-	1	1
Natural Gas Companies	3	2	2	2	2	3
Refiners, Distributors	2	2	2	4	6	4
Oilwell Transportation	-	-	5	5	8	1
Total	26	27	198	92	136	128

* Listed as Stocks and Bonds

Source: Alberta Government Telephones, Telephone Directories for Calgary and Edmonton.
A careful and thorough search was made of the Directories to obtain the number of all types of companies associated with the petroleum industry. They were then grouped into the types shown in the Table. It is believed that the Directories are highly reliable and that most companies actively involved with the industry would be listed.

obtained Alberta's second oil refinery built by the British American Oil Company. Then in 1940 the first ammonia and ammonium nitrate factory based on natural gas was constructed.

The concentration of most exploration activity in Alberta prior to 1947 in the southern part of the province made Calgary an active oil centre before the Leduc discovery. There was the gradual establishment in the city of a variety of petroleum firms and organizations which desired the amenities and services which a large centre can provide. The oil activity brought many Americans, including executives capable of directing activities, and resulted in the development within the city of a body of trained technical and managerial personnel as well as specialists in fields such as petroleum law. The oil activity contributed to the creation in Calgary of a cultural environment characterized by a willingness to invest and take risks and a strong spirit of promotion as exemplified by the annual "stampede". By 1947 the Calgary community was well oriented to oil as many residents had invested in the development of Turner Valley.

Calgary's function as an oil centre entered a phase of dramatic growth and specialization following the important post war discoveries, beginning with the Leduc-Woodbend field in 1947, and followed by a similar major finds such as the Redwater field in 1948 and the Pembina in 1953, as well as others. Companies directly involved with the various aspects of the petroleum industry mushroomed from 133 in 1946 to 2,130 in 1974.[4] Of the 2,130 firms, 1,218 represented the oil administrative group and 912 the oil operations group.

The administrative oil firms are characterized by activities broadly described as coordination and control, exchange, and the giving or receiving of particular business and technical services. They are best exemplified by the departmentalized head or branch offices of the major oil companies (producers, explorers, developers) which deal exclusively with such matters as management, finance and decision making. Other examples include the offices of independent oil operators (many of whom are small) oil brokers, geological, geophysical and engineering consultants, lease brokers, financial and investment firms, and land agents. There are also special administrative services such as oil reports and record companies, data processors as well as oil libraries. As a result of the concentration of administrative oil offices and personnel in Calgary, the city has emerged as the administrative or management centre of the

Canadian oil industry.

The large number of oil operations firms in Calgary signifies that Calgary also is an important oil operations centre. The oil operations firms are concerned with providing various services, supplies and fabricated goods for the field activities of the oil and gas industry. The field activities include ones such as seismic and oilwell drilling, oilwell servicing, oilfield construction and transportation. This group also includes the refiners, natural gas processors, and pipeline companies. Many of the oil operations firms are associated with service and repair shops, special laboratories, warehouses, garages or storage yards. Others are located in office buildings and may be concerned primarily with sales.

THE IMPACT OF THE PETROLEUM INDUSTRY ON THE CITY

The economic impact of the burgeoning Western Canada petroleum industry on Calgary after 1947 was dramatic. Within the city it created a psychological climate of buoyancy and optimism. Population growth and concomitant economic development of Calgary has been of the highest of urban centres in Canada. Population of the city more than quadrupled from 100,044 in 1946 to 403,320 in 1971.[5] A striking feature of the city's spectacular growth was the marked development of head and branch offices of administrative and oil operations firms.

The impact of the petroleum industry on Calgary is nowhere more strongly reflected than in the downtown. Large scale construction of office buildings for the oil companies and related businesses has greatly extended the downtown horizontally and has created a highly built-up, skyscraper - dominated core in which administrative oil activity is strikingly evident. Since 1960, between 600,000 and 700,000 square feet of new office space has been absorbed annually in downtown Calgary, primarily by oil firms.[6] By 1974 major new office buildings, directly or indirectly connected with oil and gas, were completed or announced for downtown Calgary with total expenditure in excess of $500 million and with office space exceeding 9 million square feet.[7]

The petroleum industry provided employment for 15,470 persons or 9 per cent of the city's labour force in 1971 with annual salaries estimated at $165 million.[8] According to the City of Calgary Planning Department, it is believed that more than 50 per cent of the city's total employment is derived from the needs of the oil industry and its employees.[9] The chairman of the Board of Governors, Canadian Petroleum Association stated that there probably were at least 10,000 office jobs directly related to the oil business in the city and that the multiplier effect, as termed by economists, creates a minimum of three and as many as seven additional jobs in business for every one in the oil business.[10]

In Calgary many new industries were created to manufacture equipment and supplies for the oil industry. Manufactured are parts and equipment for oil and gas wells, drilling rigs, gas plants, storage tanks, portable camp facilities, steel and plastic pipe for oil and gas lines as well as other items. Nearly every business and professional activity in Calgary is affected by the presence of the oil industry, from education, health, banking, manufacturing, construction, transportation and real estate to retailing, publishing and radio.

CALGARY AS A HEAD OIL OFFICE CITY

The concentration of head offices is a strong indicator of the importance of a city as an administrative centre. The head office is a company's seat of authority for it is here that its top decision makers are located. Hence it is the place from which a company's international, national, regional or local activities are directed.

A study was made of all head oil offices in Canada from 1950 to 1974 for six data years, 1950, 1955, 1960, 1965, and 1970 and 1974.[11] A cummulative total of 11,896 head offices were examined. They consisted of twelve types which were classified as belonging either to the administrative or operations group depending upon the basic function of the firm (Table 2,3). In addition, personal interviews were conducted with 112 oil offices and contact was made with numerous other persons and organizations highly knowledgeable about the development of the Canadian oil industry.[12]

The research revealed that Calgary is outstanding as a head oil office city and as such is the centre for much of the decision making affecting the Canadian oil industry. Throughout the twenty-four years from 1950 to 1974, it has been the location of approximately one-half or more of the Canadian total of head oil administrative and operations firms (Figure 2,3). During that time, its only other oil rival Edmonton has had about one-fifth of the Canadian total (Figure 2,3). Increases in number of head offices for Calgary was especially rapid after 1965 when 617 new ones were added to give the city a total of 1,582 in 1974 (Table 2,3).

Calgary is prominent particularly as a head office centre for the administrative group of oil firms. In 1950 the city had 50 per cent of the Canadian total and by 1974 had increased its lead to 75 per cent (Figure 2,3). This growth occurred at the expense of Edmonton and other centres in Canada. Increasingly these head offices are locating in Calgary.

Producers, explorers, developers, which are the key oil office type within the administrative group, grew in Calgary from 52 to 80 per cent of the Canadian head office total during the years spanning 1950 and 1974 (Table 2,3). The time of their most active development was between 1950 and 1955, when they rose in number from 230 to 396, a gain of 72 per cent. The post-Leduc oil boom was well underway by this time, oil excitment was high especially in Calgary, and many small independent Canadian companies sprang up. Oil companies from the United States also set up headquarters in Calgary. The second period of marked

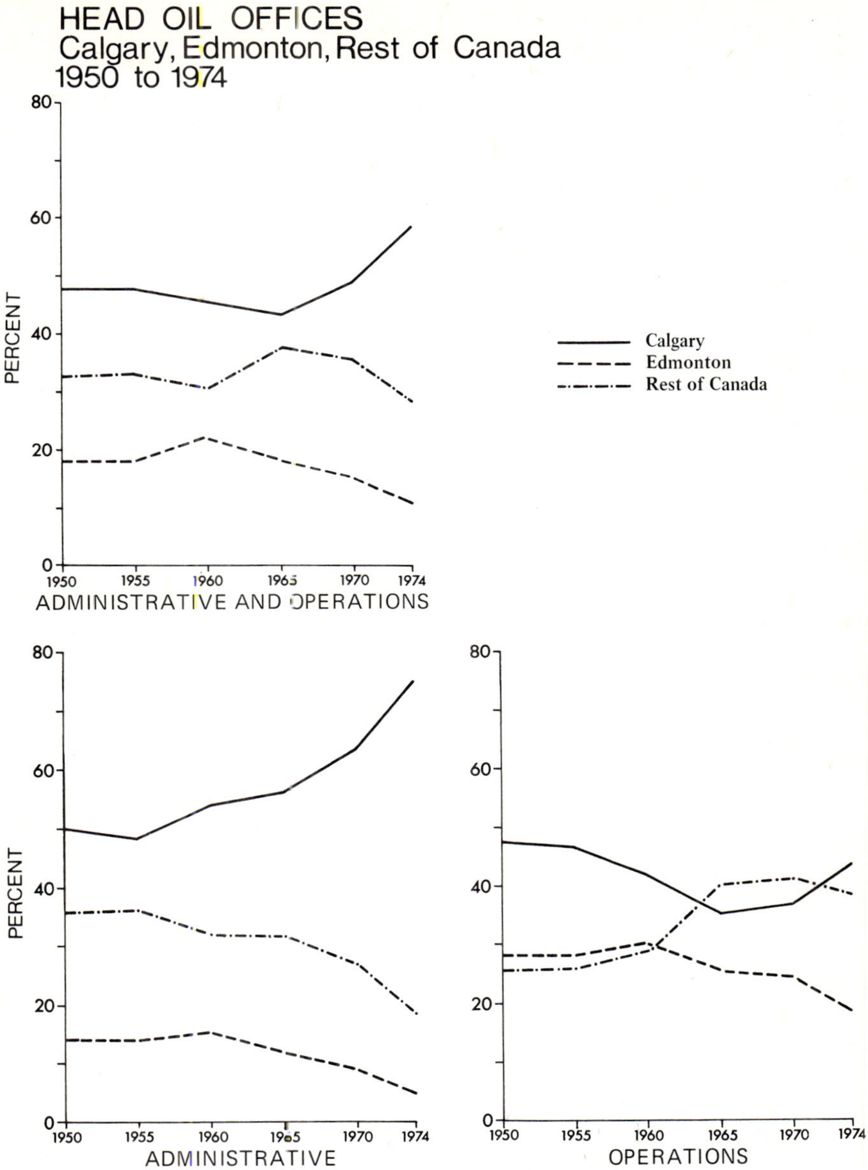

Source: Compiled from Nick e's Canadian Oil Registers

FIGURE 2,3 Head oil offices in Calgary, Edmonton and the rest of Canada, 1950-1974.

TABLE 2,3

HEAD OIL OFFICES IN CALGARY, EDMONTON, AND THE REST OF CANADA

Oil Office Group and Type	1950						1955						1960					
	Calg.		Edm.		Can.[1]		Calg.		Edm.		Can.[1]		Calg.		Edm.		Can.[1]	
	No.	%	No.	%	No.	%	No.	%	No.	%	No.	%	No.	%	No.	%	No.	%
ADMINISTRATIVE																		
Producers, Explorers, Developers	230	52	47	11	161	37	396	50	103	13	295	37	280	53	73	14	174	33
Consultants, Data Processors	25	63	13	32	2	5	38	69	8	15	9	16	88	66	24	18	21	16
Financial and Investment	18	32	1	2	38	66	12	19	4	6	47	75	17	24	6	8	50	68
Lease Brokers and Land Agents	9	37	10	42	5	21	13	68	5	26	1	6	19	58	10	30	4	12
TOTAL ADMINISTRATIVE	282	50	71	13	206	37	459	49	120	13	352	38	404	53	113	15	249	32
OPERATIONS																		
Service and Supply	92	45	61	30	52	25	122	41	95	32	83	27	185	33	196	35	184	32
Oilwell Drilling Contractors	20	44	17	38	8	18	33	51	24	37	8	12	41	65	15	24	7	11
Engineers, Designers, Constructors Fabricators	-	-	-	-	-	-	-	-	-	-	-	-	15	37	12	29	14	34
Geophysical & Exploration Drilling Contractors	59	53	28	25	24	22	94	62	29	19	28	19	91	64	23	16	28	20
Oilwell Servicing	-	-	-	-	-	-	6	32	11	58	2	10	23	37	32	51	8	12
Pipeline Companies & Power Distributors	5	36	6	43	3	21	10	50	5	25	5	25	28	47	8	14	23	39
Refiners, Processors, Marketers Plant Operators	10	45	1	5	11	50	7	22	1	3	24	75	17	34	8	16	25	50
Transportation and Oilfield Construction	-	-	-	-	-	-	-	-	-	-	-	-	-	-	-	-	-	-
TOTAL OPERATIONS	186	47	113	28	98	25	272	46	165	28	150	26	400	41	294	30	289	29
TOTAL ADMINISTRATIVE AND OPERATIONS	468	49	184	19	304	32	731	48	285	19	502	33	804	46	407	23	538	31

[1]Can. designates the number of head offices in the rest of Canada, that is, outside of Calgary and Edmonton.

Sources: Compiled from Nickle's Canadian Oil Registers.

TABLE 2,3 (continued)

HEAD OIL OFFICES IN CALGARY, EDMONTON, AND THE REST OF CANADA

Oil Office Group and Types	1965						1970						1974					
	Calg.		Edm.		Can.[1]		Calg.		Edm.		Can.[1]		Calg.		Edm.		Can.[1]	
	No.	%	No.	%	No.	%	No.	%	No.	%	No.	%	No.	%	No.	%	No.	%
ADMINISTRATIVE																		
Producers, Explorers, Developers	290	55	53	10	185	35	412	62	47	7	202	31	685	80	34	4	134	16
Consultants, Data Processors	146	66	39	18	37	16	286	73	47	12	61	15	268	72	40	11	64	17
Financial and Investment	21	25	3	4	61	71	21	27	2	3	56	70	17	25	1	1	50	74
Lease Brokers and Land Agents	37	74	8	16	5	10	55	81	5	7	8	12	42	86	2	4	5	10
TOTAL ADMINISTRATIVE	494	56	103	12	288	32	774	64	101	9	327	27	1,012	75	77	6	253	19
OPERATIONS																		
Service and Supply	227	31	199	27	313	42	286	32	233	25	396	43	321	40	167	22	308	38
Oilwell Drilling Contractors	31	69	9	20	5	11	36	71	10	20	5	9	40	82	1	2	8	16
Engineers, Designers, Constructors Fabricators	29	26	21	19	62	55	44	34	28	22	57	44	45	39	26	22	45	39
Geophysical & Exploration Drilling, Contractors	69	66	11	10	25	24	77	69	6	5	29	26	41	68	5	8	14	24
Oilwell Servicing	26	43	22	37	12	20	24	35	22	32	23	33	18	30	15	25	27	45
Pipeline Companies & Power Distributors	48	55	11	13	29	32	54	58	9	10	30	32	43	61	7	10	20	29
Refiners, Processors, Marketers Plant Operators	23	39	5	8	31	53	37	48	3	4	37	48	36	63	2	3	21	34
Transportation and Oilfield Construction	18	14	49	39	59	47	30	15	68	33	107	52	26	18	42	28	81	54
TOTAL OPERATIONS	471	35	327	25	536	40	588	36	379	23	684	41	570	42	265	19	524	39
TOTAL ADMINISTRATIVE AND OPERATIONS	965	44	430	19	824	37	1,362	48	480	17	1,011	35	1,582	59	342	12	777	29

[1] Can. designates the number of head offices in the rest of Canada, that is, outside of Calgary and Edmonton.

Sources: Compiled from Nickle's Canadian Oil Registers.

growth was 1970 to 1974 when they multiplied from 412 to 685, a jump of 66 per cent. The increase was due in part to the shift in interest and exploration activity to Northern Canada.

During this same twenty-four year time span, consultants-engineering, geological, geophysical and surveying - and data processors grew in Calgary from 63 per cent to 72 per cent of the Canadian total. Lease brokers and land agents, who are persons well versed in the law of property applying to the regions in which they work, experienced a spectacular development from 37 to 86 per cent (Table 2,3) in response to the great demand for a multiplicity of lease arrangements between oil firms and governments as well as private landowners. The only type not well represented in Calgary was head financial and investment firms of which about one half were located in Toronto.[13] Nevertheless, Calgary is the head financial centre for the petroleum industry in Western Canada. The five major banks have special oil and gas departments staffed by petroleum engineers and persons with expertise in oil industry financing, marketing, government regulations and taxation.

Calgary also is the leading head office centre for oil operations firms. From 1950 to 1974, the number of head offices of this group in the city more than tripled from 186 to 570 (Table 2,3). Calgary's 1974 total is certainly far above that for Edmonton which had only 265 head oil operations offices the same year. Calgary's share of the Canadian total, however, declined after 1950 from 47 to 35 per cent in 1965 (Figure 2,3) largely due to head offices of this group locating elsewhere in Alberta close to oil operations activities. Since 1965, Calgary's head office role for oil operations firms has strengthened as a result of a seven per cent gain. On the other hand head oil operations offices in Edmonton and the rest of Canada have diminished in number and in per cent of the Canadian total. Increasingly these offices are moving to and remaining in Calgary in order to be close to the heart of oil industry decision making.

PLATE 10
C.B.D. and residential expansion in the northwest.

City of Calgary Photo

The largest type numerically of the oil operations group is service and supply. Calgary with 321 head offices in 1974 or 40 per cent of the Canadian total, had more than Edmonton with 167 and the rest of Canada with 308 (Table 2,3). Many of the head offices which are found outside of Calgary and Edmonton in the rest of Canada were located in Alberta close to the major oil fields. It is of interest to note the numerical rise in Calgary's share of these head offices since 1970 from 286 to 321 and the sharp drop for Edmonton from 233 to 167 as well as a similar decline for the rest of Canada from 396 to 308.

Possibly the most outstanding feature of the specific types of the operations group is the dramatic rise in Calgary's share of head offices of oilwell drilling contractors, geophysical and exploration drilling contractors and also pipeline companies and power distributors. Between 1950 and 1974 Calgary's share of these three types rose 38, 15 and 25 percentage points respectively (Table 2,3). Engineers, designers, constructors and fabricators (for which data was available only since 1960) tripled in number for Calgary from 15 head offices in 1960 to 45 in 1974.

Calgary's role as an outstanding head office centre for both administrative and operations oil offices is clearly demonstrated by the results of the research. The city is growing in this function and stands as the undisputed management centre for the Canadian petroleum industry.

OIL ADMINISTRATIVE AND OPERATIONS OFFICES IN CALGARY, 1950-1974

The research of head oil offices in Calgary, Edmonton, and the rest of Canada included also a study of branch oil offices in Calgary as well as in Edmonton, the only other city with a comparatively large number. A total of 15,604 oil offices was analyzed for the six individual data years from 1950 to 1974.[14] Of this total, 7,901 were head and branch administrative and operation offices in Edmonton and the remaining 3,956 were head administrative operations offices scatterd across Canada. These figures demonstrate Calgary's dominance as a city for oil offices.

The number of head and branch oil offices for the administrative and the operations group of firms in Calgary from 1950 to 1974 for each of the six data years is shown in Figure 3,3. Growth of head and branch offices for both oil office groups throughout the twenty-four years was continuous and rapid with one exception. Head administrative oil offices had a small decline from 1955 to 1960. The graphs vividly show that throughout the entire period, head offices greatly exceeded branch offices and administrative offices outnubered operations offices. This evidence further substantiates Calgary's strong oil management function.

Table 3,3 shows the number of head and branch oil offices for the administrative and operations group of firms and for the twelve subtypes located in Calgary for the six data years. The figures for Edmonton are shown also for comparison. Throughout most of the twenty-four years from 1950 to 1974 Calgary has overwhelmingly dominated Edmonton in number of head and branch oil offices (Figure 4,3). In 1974 for example, Calgary had 1,012 head and 206 branch administrative oil offices or a total of 1,218 whereas Edmonton had 77 head and 87 branch administrative offices or a total of 164. Calgary, has had more head and branch oil operations offices with the exception that it had slightly fewer branch offices during the 1950-60 period when the Edmonton region was experiencing intense exploration and production activity accompanying the major oil strikes. Because the Edmonton community was slow to react in providing the services and supplies required on a twenty-four hour basis, particularly for oilwell drilling, many Calgary based firms as well as ones from the United States opened branch offices in Edmonton to meet the needs. For example, of the operations

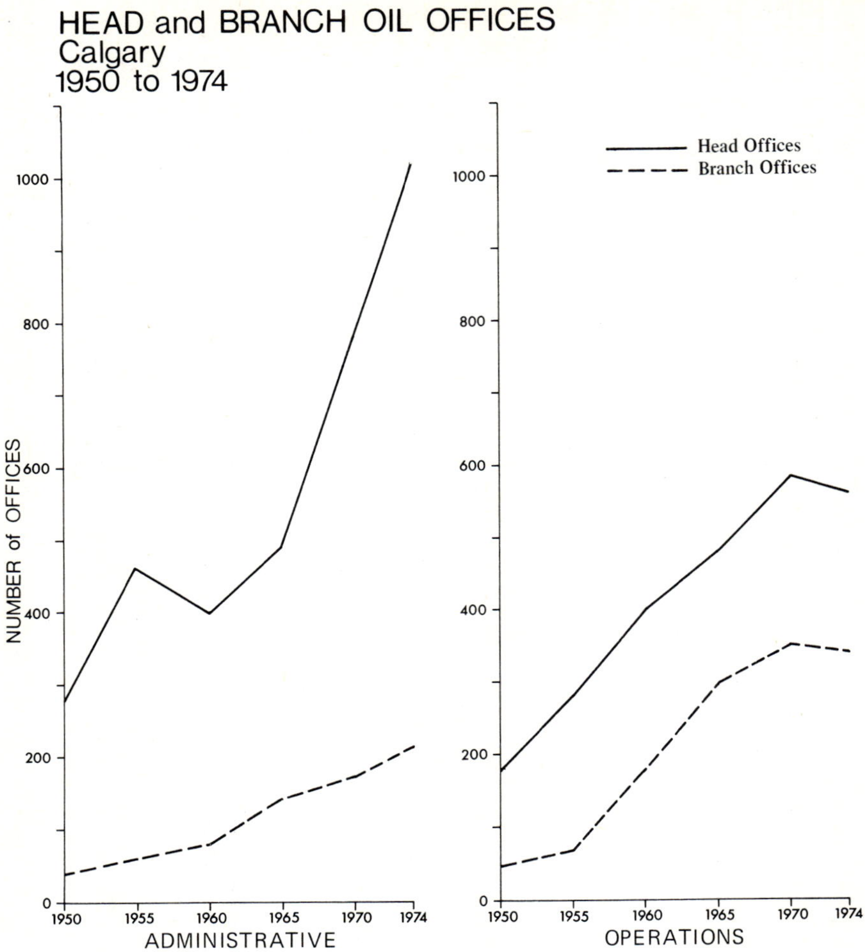

Source: Compiled from Nickle's Canadian Oil Registers

FIGURE 3,3 Head and branch oil offices, Calgary, 1950-1974.

firms with head offices in Calgary in 1950, 22 per cent had established branch offices in Edmonton, and by 1955 this figure had risen to 30 per cent.[15]

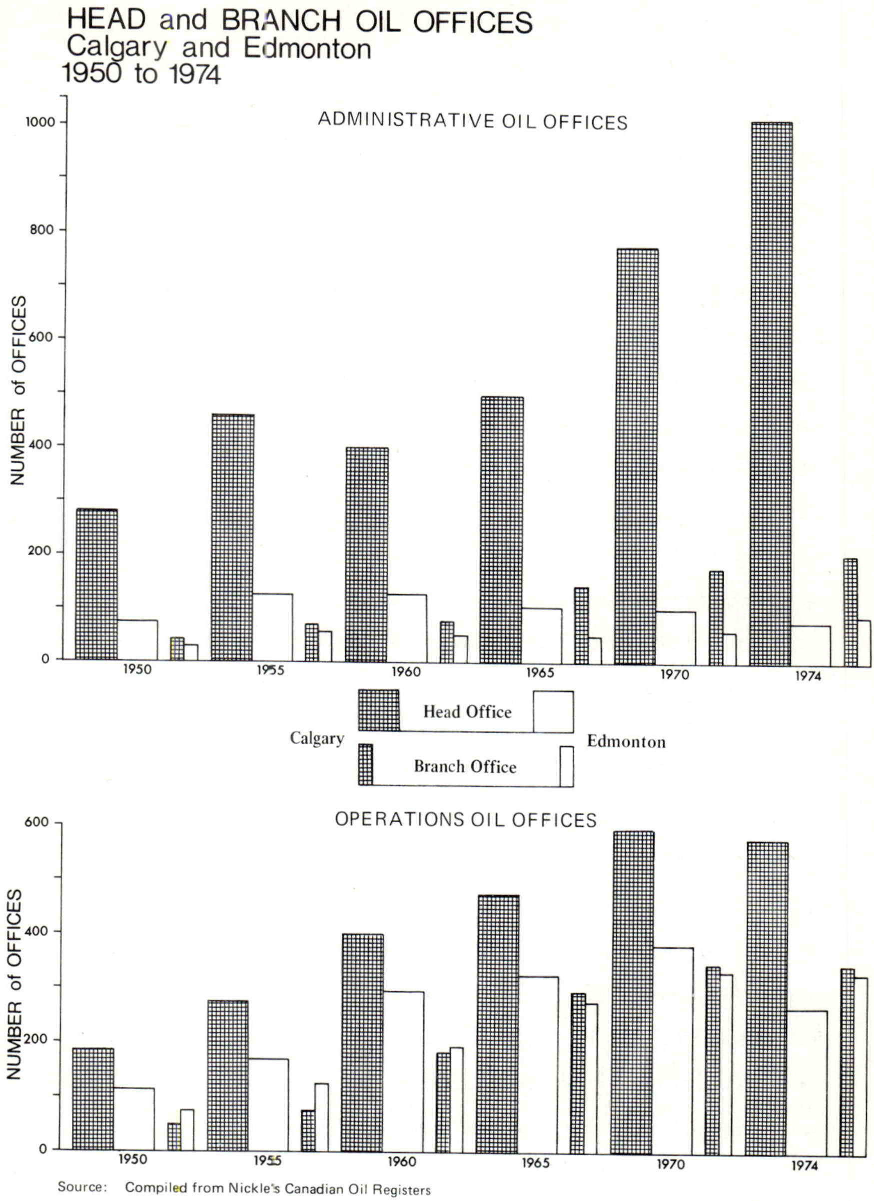

FIGURE 4,3 Head and branch oil offices in Calgary and Edmonton, 1950–1974.

Administrative Oil Offices

Producers, explorers, developers are by far the most numerous type of the administrative group having a total of 822 in 1974 (Figure 5,3) and are very much concentrated in Calgary. For instance, in 1974, Edmonton had only 75 offices of this type. The size of the companies which these offices represent vary greatly from the large multinational oil corporations such as Gulf, Shell, Texaco and Imperial Oil to very small ones owned and managed by two or three persons. Within this type there are the major integrated and major independent companies active in exploration, production and development, minor independent firms with land holdings only, and sundry royalty companies with an interest in a small number of producing wells or in mineral rights. Some of the firms of this type are split into several subsidiaries which usually are managed by the same personnel. Reasons for the division of a firm into a number of separate small companies are operational efficiency, some tax advantages, and the desire to keep different functions separate.

The time interval of most active development of producers, explorers, developers was from 1950 to 1955 (Figure 5,3) during the early post-Leduc oil boom years when they increased in number from 248 to 432. Many new companies were formed or located in Calgary in an attempt to make their fortunes in oil. Some were successful but many also went out of business. Figure 5,3 shows the drastic decline in offices of this type from 1955 to 1960. The time interval from 1960 to 1965 was one of slow increase. Sproule[16] calls this period the long five - year dry spell in major discoveries and this situation undoubtedly had a strong bearing upon the development of new oil companies and offices. The second marked increase in producers, explorers and developers, from 1965 to 1974, is related to new major oil discoveries after 1964 and to the great expansion of oil activity in Canada's North.

Consultants - geological, geophysical, engineering, surveying,

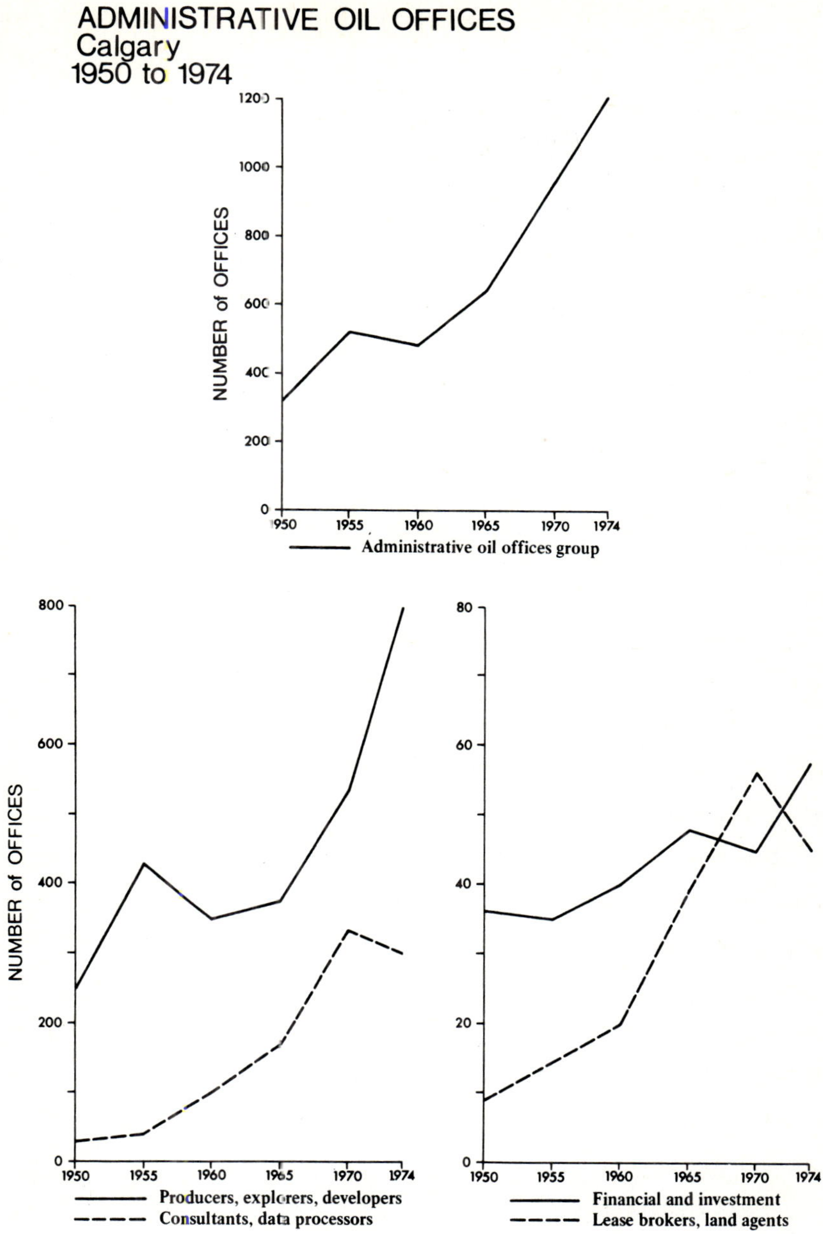

Source: Compiled from Nickle's Canadian Oil Registers

FIGURE 5,3 Administrative oil offices Calgary, 1950-1974.

and data processors are the second most numerous type of the administrative group. Their spectacular increase from 25 offices in 1950 to 319 in 1970 (Figure 5, 3) reflects the increasing demand for highly specialized technical services required by the producers, explorers, developers in the search for and development of oil and gas. The sharp numerical rise after 1965 from 169 to 319 offices is due in part to the establishment of 36 data processors in the city to process the voluminous marine and land seismic data fed into the city by Calgary based companies active in Canada and other areas such as the North Sea, the Mediterranean, Crete and Greenland. In terms of processing centres serving international petroleum explorations, Calgary ranks second behind Houston, Texas and ahead of London, England.[17] Its role as one of the three main nerve centres in the computer world is increasing as is evidenced by Shell Canada Limited's 1974 addition to its computing centre of a $4 million Univac computer rated among the biggest and most sophisticated available for scientific data processing.[18] Calgary also has the second largest known population of geophysicists and geologists.[19]

Financial and investment offices related to the oil and gas industry experienced moderate growth in Calgary from 37 firms in 1950 to 57 in 1974. This type includes the oil and gas departments of the banks, the dealers in oil stocks and bonds, investment consultants, firms assisting in oil business purchases and mergers or providing capital for oil ventures, trustees for oil royalties, and specialists in real estate and lease arrangements. In this branch of oil activity, Calgary is the leading centre for Western Canada. Supporting Calgary's position as a petroleum financial centre is the fact that major decisions in expenditures for the oil industry are made by executives based in the city.

Lease brokers and land agents are involved with activities such as petroleum and natural gas and surface leases, easement acquisitions, setting up land and lease files and records, arranging acreage deals, servicing pipeline rights-of-way, damage settlements, and client representation at crown and mining recorder sales. They have increased rapidly in Calgary especially since 1960 (Figure 5,3). This sharp rise is related to a shift of offices of this type from Edmonton to Calgary and also due to the fact that a large number of landmen who were employed by and gained experience with producers, explorers, developers during the years following 1947, started their own business.

Oil Operations Offices

Because Calgary is so outstanding in oil administration or management its important oil operations role is frequently overlooked or is little known. Public opinion recognizes Edmonton as the oil operations centre for the oil industry. The great diversity of firms and occupations represented by the oil operations group, and the widely scattered location pattern of the establishments or offices, makes the group visually less obvious than the highly clustered and imposing office towers of the administrative group in Calgary's downtown. Yet, in terms of number of oil operations offices, Calgary is the national leader with a total of 912 in 1974 compared with only 600 for Edmonton. Although there is some overlap between Calgary and Edmonton in their oil operations roles, there is a notable degree of differentiation in that Calgary's emphasis is service - sales whereas Edmonton's is oriented to service - manufacturing, transportation and oilfield construction, and oil well servicing.[20]

The most numerous and diverse members of the oil operations group are service and supply companies. They include ones such as supply houses, machine shops, tank and rig manufacturers, well testing and service companies, tool distributors, oilfield welding, repair of instruments, oilfield equipment rentals, oil well servicing and oil and natural gas well testing. Many of the firms of this type can be included in the category of light industry. However, there also are manufacturers' representatives and sales offices. Table 3,3 and Figure 6,3 show that their development in Calgary was rapid since 1950 and that in 1974 they represented more than half of the total oil operations group. If Calgary and Edmonton are compared, it is of interest to note that in 1974 Calgary had 573 firms whereas Edmonton had only 409. Calgary's increasing lead over Edmonton after 1960 is a reflection of the increasing number of branch sales offices being established in Calgary by Edmonton firms because of the desire to have close contact with the producers, explorers, developers headquartered there. Whereas in 1960 only 59 Edmonton service and supply firms had branch offices in Calgary, by 1970 this number had increased to 93.

Of the other types of oil operations offices, Calgary did numerically surpass Edmonton for the following : geophysical and exploration drilling contractors; engineers, designers, constructors and fabricators; pipeline companies and power distributors; oilwell drilling contractors; and refiners, processors, marketers and plant operators who are concerned with the refining of crude oil, the operation of natural gas processing plants for the extraction of propane, butane, natural gasoline and sulphur, and for the marketing of oil, natural gas and their products. The marked increase in firms of this type in Calgary (Figure 6,3) is strongly related to the fact that many of these firms are wholly owned by, or affiliated with producers, explorers, developers, the majority of whom are in Calgary. Calgary's increasing importance as a centre for the location of offices of pipeline companies and power distributors (Figure 6,3) also reflects the fact that many of these firms are owned or controlled by, or joint ventures of, producers, explorers, and developers in Calgary.

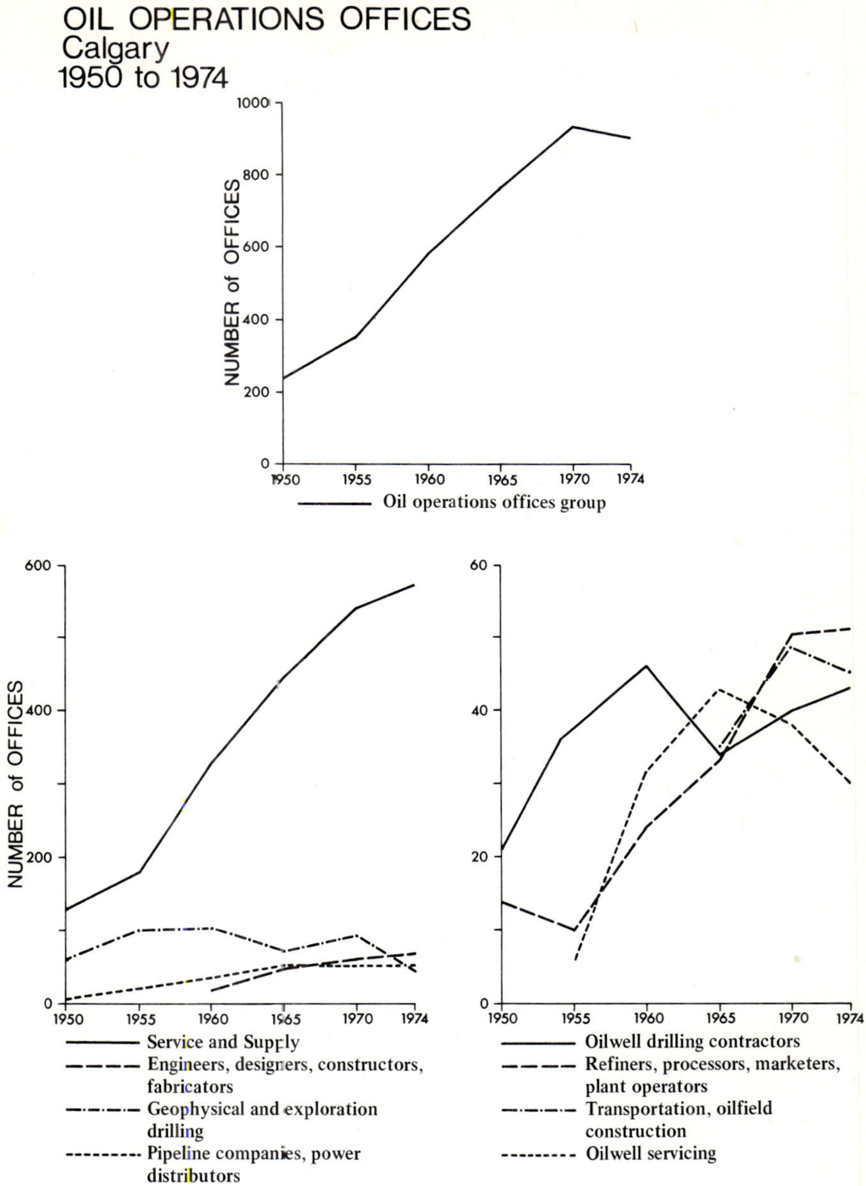

Source: Compiled from Nickle's Canadian Oil Registers

FIGURE 6,3 Oil operations offices Calgary, 1950-1974.

TABLE 3,3

OIL ADMINISTRATIVE AND OPERATIONS HEAD AND BRANCH OFFICES IN CALGARY AND EDMONTON, 1950 - 1974

Oil Office Group and Type	1950						1955						1960					
	Calgary			Edmonton			Calgary			Edmonton			Calgary			Edmonton		
	Head	Branch	Total	Head	Branch	Total	Head	Branch	Total	Head	Branch	Total	Head	Branch	Total	Head	Branch	Total
ADMINISTRATIVE																		
Producers, Explorers, Developers	230	18	248	47	8	55	396	36	432	103	34	137	280	53	333	73	22	95
Consultants, Data Processors	25	0	25	13	2	15	38	3	41	8	2	10	88	9	97	24	5	29
Financial and Investment	18	19	37	1	19	20	12	24	36	4	18	22	17	23	40	6	21	27
Lease Brokers and Land Agents	9	0	9	10	0	10	13	0	13	5	2	7	19	1	20	10	0	10
TOTAL ADMINISTRATIVE	282	37	319	71	29	100	459	63	522	120	56	176	404	86	490	113	48	161
OPERATIONS																		
Service and Supply	92	45	137	61	49	110	122	65	187	95	95	190	185	143	328	196	140	336
Oilwell Drilling Contractors	20	1	21	17	4	21	33	4	37	24	12	36	41	6	47	15	11	26
Engineers, Designers, Constructors, Fabricators	-	-	-	-	-	-	-	-	-	-	-	-	15	5	20	12	8	20
Geophysical & Exploration Drilling Contractors	59	2	61	28	4	32	94	4	98	29	4	33	91	10	101	23	9	32
Oilwell Servicing	-	-	-	-	-	-	6	1	7	11	2	13	23	9	32	32	5	37
Pipeline Companies & Power Distributors	5	0	5	6	1	7	10	0	10	5	1	6	28	3	31	8	2	10
Refiners, Processors, Marketers Plant Operators	10	4	14	1	6	7	7	3	10	1	7	8	17	6	23	8	9	17
Transportation and Oilfield Construction	-	-	-	-	-	-	-	-	-	-	-	-	-	-	-	-	-	-
TOTAL OPERATIONS	186	52	238	113	64	177	272	77	349	165	121	286	400	182	582	294	184	478
TOTAL ADMINISTRATIVE AND OPERATIONS	468	89	557	184	93	277	731	140	871	285	177	462	804	268	1072	407	232	639

Source: Compiled from Nickle's Canadian Oil Registers.

TABLE 3,3 (continued)

OIL ADMINISTRATIVE AND OPERATIONS HEAD AND BRANCH OFFICES IN CALGARY AND EDMONTON, 1950 - 1974

Oil Office Group and Type	1965						1970						1974					
	Calgary			Edmonton			Calgary			Edmonton			Calgary			Edmonton		
	Head	Branch	Total	Head	Branch	Total	Head	Branch	Total	Head	Branch	Total	Head	Branch	Total	Head	Branch	Total
ADMINISTRATIVE																		
Producers, Explorers, Developers	290	82	372	53	18	71	412	112	524	47	20	67	685	137	822	34	41	75
Consultants, Data Processors	146	23	169	39	7	46	286	33	319	47	12	59	268	28	296	40	12	52
Financial and Investment	21	26	47	3	24	27	21	24	45	2	25	27	17	40	57	1	32	33
Lease Brokers and Land Agents	37	2	39	8	0	8	55	2	57	5	0	5	42	1	43	2	2	4
TOTAL ADMINISTRATIVE	494	133	627	103	49	152	774	171	945	101	57	158	1012	206	1218	77	87	164
OPERATIONS																		
Service and Supply	227	216	443	199	205	404	286	251	537	233	236	469	321	252	573	167	242	409
Oilwell Drilling Contractors	31	3	34	9	11	20	36	4	40	10	15	25	40	2	42	1	23	24
Engineers, Designers, Constructors, Fabricators	29	23	52	21	17	38	44	27	71	28	25	53	45	28	73	26	20	46
Geophysical & Exploration Drilling Contractors	69	6	75	11	4	15	77	12	89	6	2	8	41	5	46	5	2	7
Oilwell Servicing	26	16	42	22	10	32	24	15	39	22	13	35	18	12	30	15	8	23
Pipeline Companies & Power Distributors	48	4	52	11	7	18	54	4	58	9	6	15	43	8	51	7	11	18
Refiners, Processors, Marketers Plant Operators	23	10	33	5	8	13	37	13	50	3	15	18	36	15	51	2	15	17
Transportation and Oilfield Construction	18	17	35	49	7	56	30	19	49	68	8	76	26	20	46	42	14	56
TOTAL OPERATIONS	471	295	766	327	269	596	588	345	933	379	320	699	570	342	912	265	335	600
TOTAL ADMINISTRATIVE AND OPERATIONS	965	428	1393	430	318	749	1362	516	1878	489	377	857	1582	548	2130	342	422	764

Source: Compiled from Nickle's Canadian Oil Registers.

FACTORS IN CALGARY'S FUNCTIONAL DEVELOPMENT AS AN OIL OFFICE CENTRE

To obtain data about the factors responsible for the location of oil offices in Calgary and the city's functional development as an oil office centre, personal interviews were conducted with 34 administrative and 22 operations oil offices in the city.[21] Seventy-eight per cent of the interviewees were officers, directors, owners, managers or superintendents of the firm and proved to be very knowledgeable about their company's development. Some had started their careers with the oil industry during Turner Valley days and so were able to provide valuable insights relating to oil office location decisions. Additional information was obtained from personal interviews with oil organizations such as the Canadian Petroleum Association and the Independent Petroleum Association, several presidents of oil companies, city adminstrators, as well as with other individuals associated with the petroleum industry. Interviews with oil offices in Edmonton also yielded important data.

Administrative Oil Offices

The overriding reason for administrative oil firms locating in Calgary and ranked as of most importance by producers, explorers, developers themselves was the desire to be near other firms of the oil administrative group especially producers, explorers, developers (Table 4,3). Although producers, explorers, developers are highly competitive, nevertheless they are characterized by a large number of joint ventures or shared activities which are varied in nature. Two or more companies may share the drilling of wells to reduce the high risks and costs involved, or they may form a consortium to search for oil and gas, or share a pipeline or gas processing plant. One company may "farm out" or assign part of its acreage holdings or mineral rights to another in return for a stipulated share of any oil which may be found. To negotiate the various agreements

TABLE 4,3

ADMINISTRATIVE OIL OFFICES IN CALGARY:
LOCATION FACTORS

Factors Mentioned by Interviewees	Total Weighted Responses[1]					
					All Types	
	* A	C	F	H	Total	%
Desire to be near producers, explorers, developers	60	3	12	7	82	45
Personal--historical	51	2	8	8	69	38
Desire to be near oil and gas fields or field activities	10	-	-	-	10	6
Good access to required services and supplies	8	-	-	-	8	4
Access to banks (Oil and Gas Departments)	6	-	-	-	6	3
Good air connections with other cities	3	-	-	-	3	2
Ready access to government offices[2]	2	-	-	1	3	2
Total	140	5	20	16	181	100

* A- Producers, Explorers, Developers
C- Consultants
F- Financial, Investment (Oil and Gas Departments of banks)
H-Lease Brokers Land Agents

1 Each interviewee was asked to state in order of importance the two or three most important factors considered by the firm in its location decision. The first mentioned factor was given a weight of three, the second two and the third one. The weights for each factor were totalled according to the number of responses to obtain the total weighted responses.

2 Provincial Government offices eg. Oil and Gas Conservation Board, Land Titles Office, and Court House.

Source: Personal interviews, 1970

necessitates that companies get together frequently on a face-to-face basis. Therefore, a location in the same city offers great benefits in terms of time, covenience and accessibility. For example, between 1941 and 1946 the McCall-Frontenac Oil Company (Texaco) participated with Union Oil of California in the drilling of oil wells in order to reduce the high risks and cost involved. Similar joint ventures are constantly being negotiated. The need for oil firms to meet frequently was reinforced after 1947 by the introduction of oil field unitization. Field unitization is an arrangement whereby the wells in an oil field, which may be owned by as many as twenty or thirty companies, are operated under a common management which permits the application of good engineering practices to the entire field. Managerial and technical personnel from the various oil companies must meet to prove their share of production and draw up agreements whereby the oil pool is operated by one company or by a chosen committee. In 1957 the Alberta Government amended the Oil and Gas Conservation Act to provide for compulsory unitization when requested by owners of more than 50 per cent of the interest in an oil pool and if deemed in the public interest. Some companies have personnel who are constantly involved with unitization meetings so the advantages of being together in Calgary are apparent.

The desire to be near producers, explorers, developers was also the major factor in the concentration of lease brokers and land agents, and consultant and data processors in Calgary. Many of the producers, explorers, developers are not large enough to have their own technical consultants, land personnel or computing facilities so must depend on outside professional help. Lease brokers, land agents and consultants have located in Calgary to meet this need.

The second most important factor in the location of administrative oil offices in Calgary and in the city's development as the Canadian oil administrative centre was personal - historical (Table 4,3). This factor

PLATE 11
C.B.D. looking west. Older brick structures have been largely replaced in the past fifteen years.

Knowlton Realty Photo

includes reasons such as the early start with the oil industry, entrepreneurs gaining petroleum industry experience in the city's pre-Leduc oil climate, the location of a predecessor company in the city, the personal contacts that oil industry personnel had with oilmen already in Calgary, a personal preference for the business and social climate of Calgary, and good access to the recreational facilities of the Calgary region.

When the Leduc oil discovery was made near Edmonton, Calgary already had a substantial number of oil companies of the producers, explorers, developers type as well as other experienced personnel such as geologists, geophysicists, land agents and oil brokers because of Turner Valley and the early exploration activity in Southern Alberta. For instance, in 1919 Imperial Oil located in Calgary and in 1921 its subsidiary Royalite acquired the locally born Calgary Petroleum Products Company which had made the initial Turner Valley discovery in 1914.[21] In 1934 the British American Oil Company established themselves in Calgary by purchasing the Bell Refining Company in the city.[22] Imperial Oil and the British American Oil Company contributed funds to R.A. Brown of Calgary for the drilling of the important 1936 discovery well in the Turner Valley Field. Shell Oil had two persons in Calgary in 1941 holding leases.[23] In 1944, two geologists from Socony-Vacuum Exploration Company (now Mobil Oil Canada Limited) opened an office in Calgary because they knew Imperial Oil and Shell personnel.[24] The Superior Oil Company (now Canadian Superior Oil Limited) had geologists in Calgary and Edmonton from 1943 to 1946 and because of contacts in Calgary they recommended that an office be opened there; this was done in 1946.[25] Calgary was the location of independent oil companies such as Home Oil established in 1929 and the British Dominion Oil and Development Corporation (now Asamera Oil Corporation) both of which were started by Calgarians. Similarly, other companies located in Calgary because producers, explorers, developers already were there and because of previous contacts with persons in the city.

Concerning Calgary's early start, Hoover has stated that "a locality with an early start in some industry has thereby a complete advantage that it may retain and increase even though the early start was due to pure chance or whim." [26]

Of import in the location of administrative oil firms in Calgary was the fact that petroleum clubs and organizations were first established there so oil people coming into Alberta would go to Calgary to make contacts and to find out what was happening in oil activity. These oil people got to know each other in Calgary, fraternized through the petroleum clubs, and became acquainted with the local residents; business and social ties became well established. Therefore, when oil was discovered near Edmonton at Leduc in 1947, the oil firms already had their roots fixed in Calgary and rather than moving head offices to Edmonton they opened branch offices instead or persons travelled there if the need required.

The physical and cultural environment of Southern Alberta and the spirit of Calgary characterized by ranching, the cowboy, and the annual "stampede" appealed to the oilmen because it was similar to what they had known at home in Texas, Wyoming and Montana. Calgary's proximity to the Rocky Mountains and to Banff National Park for recreation also were an attraction. Calgary was much better known in the United States than Edmonton because of the "stampede" and Banff.[27] In this regard, a study by the University of Michigan, Institute of Social Research has found that preferences for climate and recreational facilities were important location factors in the choice of a specific town.[28]

The exercise of the various aspects of administration in an oil region requires office accommodation. Weber has stated that the city which "provides the greatest amount of office space may find itself in an excellent position to take advantage of the administrative opportunity promised by the oil industry" and " once the necessary accommodations

PLATE 12
The C.B.D., and Institutional foci
in northwestern Calgary.

Knowlton Realty Photo

are provided administrative establishments and personnel thrive and multiply in the area provided the area continues profitable in its location."[29] Following the major oil discovery in 1947, many new oil companies came to Alberta and others expanded so that for several years there was a severe shortage of office space in both Calgary and Edmonton. Any small space could be rented as an office. The Calgary community in keeping with its western spirit was quick to react. Money was invested in new office buildings and unused storage and retail space and older loft warehouses were converted to office use. The Canadian Pacific Railway's Palliser Hotel provided the first office space for numerous oil firms. Therefore, during the critical years from 1947 to 1960 when both Calgary and Edmonton were experiencing the greatest influx of oil firms, Calgary with its characteristic strong spirit of promotion did everything possible to accommodate and attract the administrative oil offices, and in this it succeeded. Saarinen[30] points out that between 1953 and 1962 office space in Calgary's central business district almost doubled from 24.5 per cent of the total space to 44.4 per cent.

Calgary's proximity to the United States and the American influence in Calgary were also significant in its development as the oil administrative centre. Most Canadians look upon Calgary as an Americanized city because of the large number of American citizens who have made it their home.[31] Some came over the trails from the United States with the advent of the ranching industry and with them came the spirit characteristic of the range-- the friendliness and a willingness to invest and take risks. More came with the discovery of oil and gas at Turner Valley and the early exploration activity in Southern Alberta. Calgary was not only the largest centre close to Turner Valley but also to the United States. Therefore the American geologists, geophysicists, drillers, and the executives in coming to Alberta located in Calgary which could provide the amenities desired and where a well established group of Americans already resided.

Furthermore, Calgary was closer to the Montana and other American oil-fields than Edmonton.

Another reason given by administrative oil firms for choosing a Calgary location was the desire to be near oil and gas fields or to field activities. This factor reflects in particular the exploration activity in Southern Alberta prior to 1947.

Some oil administrative oil firms chose to locate in Calgary because of the good access to required services and supplies. Calgary's early start as a service and supply centre for the petroleum industry because of Turner Valley and its numerical dominance in firms of this type today has already been discussed. A Calgary location offers the administrative firms, especially the producers, explorers, developers the advantage of ready access to the diversity of service and supply firms.

The factor, access to banks, is significant because of the location in Calgary of the oil and gas departments for each of the five major banks. The well qualified oil technical staff are able to provide the expertise in planning oil financing programmes for all phases of the petroleum industry. Four of the five banks had either an oil and gas department or an oil and gas manager prior to 1956.

Transportation was a factor in the development of Calgary as the administrative centre. Many of the administrative oil firms which located in Calgary were branch offices of United States firms so by 1947 there existed a well established pattern of oil personnel movement between Calgary and American oil cities. Even during the 1950's when the location pattern of oil offices was becoming firmly established for Calgary and Edmonton, travel by road was still a major undertaking and Calgary was 200 miles closer to the United States than Edmonton. Ability to get not only to oil centres such as Dallas, Houston, Tulsa but also to the money markets of New York was better by rail and air from Calgary than Edmonton.[33] Calgary rather than Edmonton was therefore the choice in the location of new administrative oil offices by American firms.

Oil Operations Offices

Research of the factors in the location of oil operations firms in Calgary revealed that the dominant one was the desire to be near producers, explorers, developers. Other important factors included: personal-historical; desire to be near oil and gas fields or field activities; and the good air connections with other cities. Of lesser importance were: access to the oil and gas departments of the banks; access to other operations offices; and proximity to the United States (Table 5,3).

The strong desire by oil operations offices to be close to the producers, explorers, developers, the majority of whom are in Calgary, is because of the need for personal contacts with them to obtain sales and work. The purchasing and engineering departments of the producers, explorers, developers are centralized in Calgary so many service and supply firms have located their head or branch sales office in the city and are doing so increasingly. Earlier, reference was made to the increasing number of service and supply firms with head offices in Edmonton which are establishing branch offices in Calgary. Also, small service and supply companies which cannot maintain offices in the two cities because of the expense choose to locate in Calgary to be near the producers, explorers, developers for business. The necessity of face-to-face contact with oil companies has brought to Calgary many geophysical and exploration drilling and oilwell drilling contractors and has contributed to Calgary's numerical lead over Edmonton in offices of this type. Also, certain operations offices which provide specialized services for oil companies such as data s-orage located in Calgary in order to have direct contact and to give rapid service.

The personal-historical factor in the location of oil oil operations offices in Calgary relates particularly to the oil and gas discoveries and exploration activity in Southern Alberta prior to 1947. Some operations offices were started in Calgary by Americans and some by local people

TABLE 5,3

OIL OPERATIONS OFFICES IN CALGARY: LOCATION FACTORS

Factors Mentioned by Interviewees	Total Weighted Responses[1]								
	B*	I	L	E	G	D	J	All Types Total	%
Desire to be near producers, explorers, developers	30	3	3	3	6	6	12	63	52
Personal-historical	14	-	-	-	2	-	2	18	15
Desire to be near oil & gas fields or field activities	4	-	2	-	2	-	6	14	11
Good air connections with other cities	10	-	-	-	-	2	-	12	10
Access to banks (Oil & Gas Departments)	2	-	-	-	1	4	-	7	6
Access to other operations offices	2	-	1	2	-	-	-	5	4
Proximity to the United States	1	2	-	-	-	-	-	3	2
Total	63	5	6	5	11	12	20	122	100

*

B- Service and Supply

I- Oilwell Servicing

L- Transportation and Oilfield Construction

E- Engineers, Designers, Constructors, Fabricators

G- Geophysical and Exploration Drilling Contractors

D- Oilwell Drilling Contractors

J- Pipeline Companies and Power Distributors

1

Each interviewee was asked to state in order of importance the two or three most important factors considered by the firm in its location decision. The first mentioned factor was given a weight of three, the second two and the third one. The weights for each factor were totalled according to the number of responses to obtain the total weighted responses.

Source: Personal interviews, 1970.

to serve the Turner Valley and other oil activity prior to Leduc. Of some significance as a personal factor was the preference by Americans for proximity to the recreational facilities and ranching areas of the Calgary region. Also cited as an attraction was the flamboyant spirit of Calgary as typified by its renowned "stampede".

The third location factor, the desire to be near oil and gas fields or field activities, reflects Calgary's proximity to the oil and gas fields discovered in its region before Leduc as well as those found in Central Alberta after 1947. Greatly improved transportation has given Calgary based operations firms relative ease of access to many of the post-Leduc oil and gas fields.

The last major factor, good air connections with other cities, refers to Calgary's relatively good air access to oil centres in the United States such as Houston and Tulsa. For example, over forty petroleum equipment suppliers and related businesses in Calgary are associated with petroleum firms in Houston[34] and there is considerable movement of personnel between the two centres.[35]

TRENDS IN CALGARY'S GROWTH AS AN OIL CENTRE

Administrative Oil Offices

Calgary is strengthening its position in oil administration. Since about 1960 there has been a marked shift of administrative oil offices to Calgary from Edmonton as well as from other parts of Canada and this trend appears to be continuing. Administrative oil offices in Calgary increased from 490 in 1960 to 1,218 in 1974 whereas Edmonton witnessed a decline from 161 to 158 in 1970 and then a small increase to 164 in 1974 (Table 3,3).

The spectacular development of producers, explorers, developers in Calgary from 333 in 1960 to 822 in 1974 and the decline in Edmonton from 95 to 75 reflects in part a shift of both head and branch offices from Edmonton to the southern city. This shift of offices has been accompanied by a reduction in size of some Edmonton producers, explorers, developers offices with the centralization in Calgary of personnel involved with decision making (that is managers, geologists, geophysicists and engineers). For instance, in 1971 Shell moved its 300 - employee Exploration Production and Land Division office from Edmonton to Calgary. The same year the D.M. Vipond Company, a small producer, also moved. In 1972 Imperial Oil, the only very large office of the administrative type in Edmonton, moved its 200 - employee Production Department to Calgary. Factors responsible for this trend are the need for face-to-face contact with staff of other administrative oil offices, a desire to have decision-making personnel together, the need for greater efficiency and cost reduction, the increasing spread of oil company activities to all parts of Canada and improvements in communication and transportation so that field activities can be reached almost as easily from Calgary as Edmonton or elsewhere. In addition, major oil companies are moving professional personnel to Calgary to get them closer to better computer facilities, to the computer banks of geological data, and in closer touch with technical experts involved in developing Canada's North. Almost all of the major consortiums searching for new oil reserves in the Arctic and studying northern oil and gas pipeline technology are based in Calgary. Included among these is Panarctic Oils Limited which is backed by the government of Canada and by a consortium of Canadian oil companies.

Lease brokers and land agent offices also have moved from Edmonton to Calgary. The dramatic increase in offices of this type in Calgary and the outstanding decline in Edmonton since 1960 (Table 3,3) has already been mentioned. Although there are some advantages in being close to the Provincial Government offices in Edmonton, nevertheless, of greater importance is the need for direct contact with producers, explorers, developers in Calgary for business in order to survive.

Oil Operations Offices

The most marked trend concerning oil operations offices has been the establishment and centralization in Calgary of sales offices and personnel by service and supply firms because of the need for close contact with producers, explorers, developers for sales and business. The sharp rise in Calgary's service and supply firms especially after 1960 is due not only to new firms locating in the city but also to firms with head offices in Edmonton and elsewhere opening branch offices. For instance, the number of Edmonton service and supply head offices with branch offices in Calgary increased from 59 in 1960 to 76 in 1974. Between 1970 and 1974, Calgary gained 36 service and supply offices whereas Edmonton lost 60 (Table 3,3).

The time period 1947 to 1960 was the one during which Edmonton had its greatest number of producer, explorer, developer companies. Because of the intense oil activity in the Edmonton region during that time, the majority of the business for service and supply firms was obtained from these Edmonton firms. Now that exploration activity in the Edmonton region has declined greatly and most of the producers, explorers, developers have centralized their purchasing and engineering departments in Calgary, many of the service and supply firms from Edmonton and elsewhere have found it necessary to move sales and management personnel to Calgary for the important sales contacts.

Another reason for the shift of service and supply as well as some of the other oil operations firms to Calgary is that producers, explorers, developers have changed to bulk buying of supplies and oilfield services by means of a bidding system. Producers, explorers, developers state the need for a specified allotment of supplies or services for as much as an entire year and then oil operations firms submit bids. Proximity to and personal contact with producers, explorers, developers to obtain work or contracts for supplies has become of prime importance to many oil operations firms. Finally the slow down in some aspects of oil industry activity since about 1965 has intensified the competition for business and so has contributed to the desire for a Calgary location.

One area of oil operations in which Calgary is declining is in refining. Of Calgary's two refineries, the Gulf Oil refinery (formerly British American Oil Company refinery) has been converted to manufacture only asphalt and the refining operation has been moved to Edmonton. Imperial Oil's refinery was to have been closed in 1974 upon completion of the company's new Edmonton refinery. This change in no way poses a threat to Calgary's distinct oil operations service-sales role which actually is increasing.

PROSPECTS FOR THE FUTURE

Calgary should continue to thrive as the administrative centre for the Canadian petroleum industry. Increasingly oil industry decision making is being concentrated in the city not only from Edmonton but also from other Canadian centres. Even though Calgary's advantages for oil office location should decline in future, the inertia of invested capital should help to perpetuate the city's strong oil management function. Weber states that "major oil centres that have attracted a more permanent share of the industry usually have kept this share. Tulsa keeps its oil offices, although the local boom that created Tulsa has since expired."[36]

In spite of the fact that oil activity is spreading out and becoming national in scope, Calgary still is the base for the directing of activities. Improved air transportation is a very important factor in this development. Indeed, one major producer, explorer, developer aptly pointed out that today it is just as easy to direct exploration activity in Eastern Canada from Calgary as from Toronto, and a location in Calgary has the added advantage of being close to other oil companies and decision makers.[37]

The immediate future for Calgary as an oil office centre appears good. The Alberta Oil and Gas Conservation Board in 1968 estimated that proven remaining reserves of crude oil and natural gas were adequate to serve Canadian and American markets for over thirty years at the then current rates of consumption.[38] Furthermore, the McMurray tar sands in Northern Alberta with their vast oil potential are just beginning to be tapped for oil. Also, Calgary as the head office centre for the Canadian oil industry will undoubtedly benefit from the oil and gas discoveries in Northern Canada.

REFERENCES

1. Telephone communication. Public Relations Department, Imperial Oil Limited, Edmonton, April 7, 1971. Turner Valley became the chief source of crude oil for the refinery after 1936.

2. HANSON, E.J. Dynamic Decade: The Evolution and Effects of The Oil Industry in Alberta. Toronto : McClelland and Stewart Limited, 1958, p. 49.

3. Prorationing in simplest terms is just a system of regulating oil production. In 1938 the Alberta Government set up a Conservation Board to regulate the output of each producer and of each of his wells.

   The way in which prorationing now works is well outlined by Gray as follows:

   > Refiners wishing to purchase Alberta oil must nominate each month with the Conservation Board for the amount of oil they require. The total of these nominations represents the market demand for any particular month. The market demand is then prorated among the province's oil fields in proportion to the estimated recoverable oil reserves for each field. In similar fashion, the resulting total for each field is prorated among the wells within that field...
   >
   > Under Alberta's prorationing plan, the share of the market demand which any producer receives is not dependent upon the number of oil wells which he owns, but upon the estimate of his recoverable oil reserves. GRAY, E. Impact of Oil. Toronto: The Ryerson Press and Maclean-Hunter Limited, 1969, pp. 33-34.

4. The 1946 figure was compiled from the 1946 Alberta Government Telephone Directory for Calgary and the 1974 figure from Nickle's Canadian Oil Register, 1973-74. Calgary : C.O. Nickle Publishing Company.

5. The population figures are for metropolitan Calgary. The 1946 figure is from Calgary Power Limited. Alberta Province of Opportunity: A Survey of Resources and Economic Possibilities. Calgary: J.T. Donald and Company Limited 1958; the 1971 figure is from: 1971 Census of Canada, Population, Special

Bulletin Feb. 1973.

6. City of Calgary, Incustrial Development Department, Onward, Vol. 5, No. 4, July - August, 1974, p.2. Summary of an address to the Rotary Club of Calgary by Mr. G. Knowlton, president of Knowlton Realty, largest office developer and leasing agent in Calgary.

7. Figures were obtained from a careful tabulation of all oil related office tower construction completed and announced up to January 1974.

8. City of Calgary, Industrial Development Department, Onward, Vol. 5, No.2, March - April, 1974.

9. City of Calgary, Planning Department, Future Labour Force, Calgary, 1966 - 1986. Calgary Metropolitan Area Environmental and Transportation Study 1966, p. 12.

10. GROSSMAN, W.L. "Calgary Remains a Gas, Oil Centre?" in Onward Calgary, Vol. 3, No.2, March - April, 1972, P.3. Mr. Grossman was the 1971 chairman of the Board of Governors, Canadian Petroleum Association and is vice-president of Shell Canada Limited.

11. See ZIEBER, G.H. Inter-and Intra-City Location Patterns of Oil Offices for Calgary and Edmonton, 1950-1970. Unpublished Ph.D. Thesis, University of Alberta, Edmonton, 1971. The statistics from this study were updated to 1974. The basic source for the data was Nickle's Canadian Oil Registers, Calgary, C.O. Nickle Publishing Company (Published annually).

12. Of the 112 persons interviewed, 78 per cent were executives. They included officers, directors, owners, managers and superintendents. Some of the other persons and organizations interviewec were the Canadian Petroleum Association, the Independent Petroleum Association, Mr. S. Raborn, president of Canadian Delhi Oil Limited and member of the Board of Directors of Calgary Inn, Mr. E.H. LaBorde, president of Laborde Petroleums Limited, and Dr. E. Roper, former mayor of Edmonton.

13. ZIEBER, op. cit., Appendix A, p. 255.

14. Although a total of 15,604 oil offices were analyzed for the six data years from 1950 to 1974, it must be noted that the actual number which existed was less because the same office may have been present during several or all of the six data years.

15. ZIEBER, op. cit., Table 19, p. 146.

16. SPROULE, J.C. "Exploration and Discovery," in HILBORN, J.D. ed., Dusters and Gushers : The Canadian Oil and Gas Industry. Toronto : Pitt, 1968, p.18

17. City of Calgary, Industrial Development Department, Onward, Vol. 4, No. 2, March - April, 1973, p.3.

18. City of Calgary, Industrial Development Department, Onward, Vol. 5, No. 2, March - April, 1974, p.2.

19. City of Calgary, Industrial Development Department, Onward, Vol. 4, No. 2, March - April, 1973, p.3.

20. ZIEBER, G.H. "The Dispersed City Hypothesis with Reference to Calgary and Edmonton," The Albertan Geographer, No.9, 1973, Edmonton : Department of Geography, The University of Alberta, pp. 4 - 13.

21. The sampling method used to select the oil offices to be interviewed was "purposive or judgement" sampling (see HARVEY, D. Explanation in Geography. London, Edward Arnold, 1969, pp. 359-361). Emphasis was given to those major firms, especially producers, explorers, developers, which located in Calgary prior to the 1950's for it was assumed that these offices established the basic location patterns and would have a marked bearing upon firms which were attracted to the city in the years to follow.

22. Personal interview, W.E.Smith , Regional administrative Assistant, Imperial Oil Limited, Calgary, July 1970.

23. Gulf Oil Canada Limited, Gulf Commentator, Toronto, Vol. 28, January 1969, p.23.

24. Personal interview, Shell Canada Limited, Calgary, July 1970.

25. Personal interview, B. Taylor, Financial Vice-President, Mobil

Oil Canada Limited, Calgary, July 1970.

26. Personal interview, J. Pyle, Treasurer and Controller, Canadian Superior Oil Limited, Calgary, July 1970.

27. HOOVER, E.M. The Location of Economic Activity. Toronto : McGraw-Hill Book Company, 1948, p. 4.

28. Personal interview, E. LaBorde, President, LaBorde Petroleums Limited, Calgary August 1970.

29. Survey Research Center, Institute of Social Research, University of Michigan, Industrial Mobility in Michigan. University of Michigan, May 1951, p.2.

30. WEBER, D. A Comparison of Two Oil City Business Centres (Odessa-Midland, Texas). Chicago : Department of Geography Research Paper No. 60, University of Chicago, 1958, pp. 153, 159.

31. SAARINEN, T.F. The Changing Office Functions in Calgary's Central Business District 1942-62. Unpublished M.A. Thesis, Department of Geography University of Chicago, 1963, p.58.

32. LIDDELL, K. Alberta Revisited. Toronto: The Ryerson Press, 1960, p. 125. Liddell reports that there are some 30,000 American citizens residing in the area served by the United States consul's office in Calgary.

33. Personal interview, E. LaBorde, op. cit.

34. Aviation and Urban Transportation Department, Houston Chamber of Commerce, Houston's Community of Interest with Other Major Oil Producing Areas. Houston, June 1970.

35. Personal interviews with oil operations offices, Calgary, 1970.

36. WEBER, D. op. cit, p. 222.

37. Personal interview, Shell Canada Limited, Calgary 1970.

38. Alberta Government, Bureau of Statistics, Department of Tourism. Alberta Industry and Resources, 1970 Edition. Edmonton : Queen's Printer, 1970, p.66.

PLATE 13
High-rise apartment buildings, the western end of the C.B.D.

Knowlton Realty Photo

CHAPTER 4

CHANGES IN THE CENTRAL PLACE SYSTEM AROUND CALGARY: 1951 – 1971[1]

Wayne K.D. Davies
and
T.T. Gyuse

University of Calgary

INTRODUCTION

Two rather diverse bodies of literature, usually summarized as the 'central place'[2] and 'growth pole' approaches[3], constitute the bulk of our present knowledge about the structure of the inter-urban system. However, both approaches contain serious restrictions that impede their overall utility. In the case of the central place literature the static nature of the theory and the limited range of empirical evidence about the dynamics of change constitute one major problem. In addition the effect of economies of scale and agglomeration economies in urban or central place growth are both ignored. Hence the effect of metropolitan inspired change is discounted, although Lösch's[4] introduction of a central city to organize his market networks constitutes an important modification of the original theory. Fortunately for our understanding of the nature of the inter-urban system, some, although far from all, of these problems have been resolved by the development of a series of growth pole concepts. But so far these concepts have not been articulated into a viable theory, and there still remains a great deal of ambivalence about the respective utility of such alternative mechanisms as 'polarization', or 'spread and spin off' processes. By the former process growth around large cities is inhibited, because people, jobs, and shops are attracted to the centre. Hodge[5] has provided an example of this restriction of growth around the larger centres in Saskatchewan. In the latter, the growth of small towns around the central core is stimulated by the 'spinning off' of functions from the metropolitan centre. The result, as Pahl[6] has shown, is growth within the metropolitan hinterland, and the development of a series of specialized centres to replace the hierarchical pattern postulated by the traditional central place approach. In recent studies of East Anglia and Brittany, Moseley[7] has warned that the 'spin off' or 'spread' effects are quite limited, so the appropriate empirical relevance of the growth pole trends is still in some doubt. In any case it is quite likely that 'polarization' and 'spin off or spread' processes may be more relevant for certain city sizes, or economies, or even growth indicators. Moreover, they may be present in different degrees in various sectors of any single metropolitan hinterland, perhaps in the same way that sectoral, concentric and nuclei trends were unravelled by Haggett[8] in a study of growth around Cambridge, England. Hence one must be wary of too rigid an application of either the 'spread' or 'polarization' mechanisms and the rejection of central place processes. The empirical evidence about growth poles is still rather limited since it is derived from case studies of very different economies in which a variety of indicators of change, from population to commercial activity and industry, have been used. So we cannot assume the presence of an inexorable set of trends having universal relevance.

This study is designed to contribute to the debate on these problems in the growth pole and central place literature by focusing on the nature of change in metropolitan hinterlands. It is, however, restricted to one of the indicators of change, that of the commercial sector, and to settlements in the metropolitan area of Calgary during the

PLATE 14
Before 1960, The Palister Hotel was the C.B.D.'s most prominent building.

City of Calgary Photo

period 1951-1971. The general objective of the essay is to determine the functional or commercial structure of the system of settlements in the area for the two years, and to comment on the changes that have taken place. More specifically the essay seeks to determine the utility of the hierarchy postulate of traditional central place theory in understanding the pattern of places and functions. It also investigates whether the changes that occurred between 1951 and 1971 show the 'upward shift' of functions and 'squeezing out' of the middle range of places described by Berry[9] and Stabler and Williams.[10] As Calgary is the fastest growing metropolitan area of Canada, one might also expect modification of the original hierarchical pattern by one or more of the growth pole processes emanating from the city, either the growth inhibition described by Hodge in Saskatchewan,[11] or the development of a series of specialised centres by 'spread' mechanisms. But before dealing with these issues a review of the area, data, and techniques used in the study is needed.

STUDY AREA, DATA AND TECHNIQUES

The metropolitan area defined for the purpose of this analysis consisted of the region within eighty miles of Calgary. This is more than the one hour travel time zone often used to define the area of regular metropolitan contact, but seemed justified by two factors. The first is the size of the area in which the two Calgary daily newspapers dominated their competiters from Edmonton, Red Deer and Lethbridge, (Figure 1,4). The second is that the choice of such a generously defined area avoided the possibility of any boundary resolution problems corrupting the results. Within this area, Figure 2,4 shows that there is a major belt of population concentration running north-south through Calgary roughly aligned along the

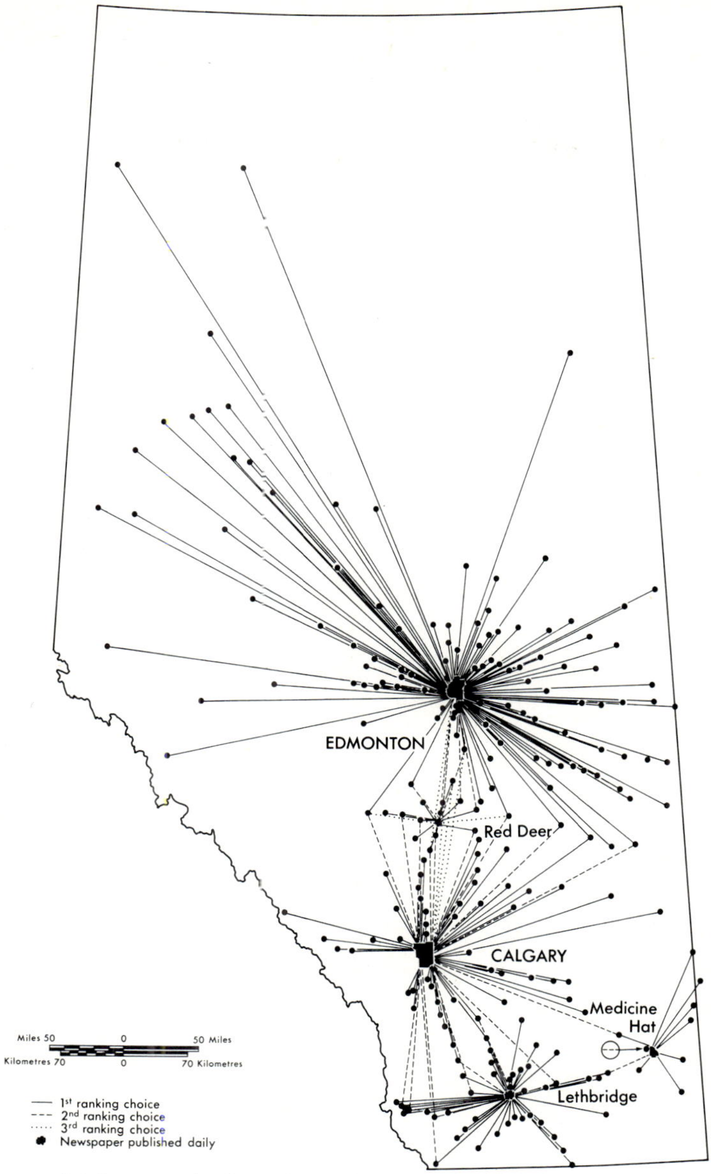

Desire lines connect each settlement with centers publishing daily newspapers. From the total sales of daily newspapers derived from the Audit Bureau of Circulation, first and subsequent rankings of the most frequently read daily newspaper in each settlement are made. However, lines are only shown for daily papers selling over 10%. Papers with sales under 10% of total number of daily newspapers are not shown.

FIGURE 1,4 Daily newspaper circulation in Alberta, 1971.

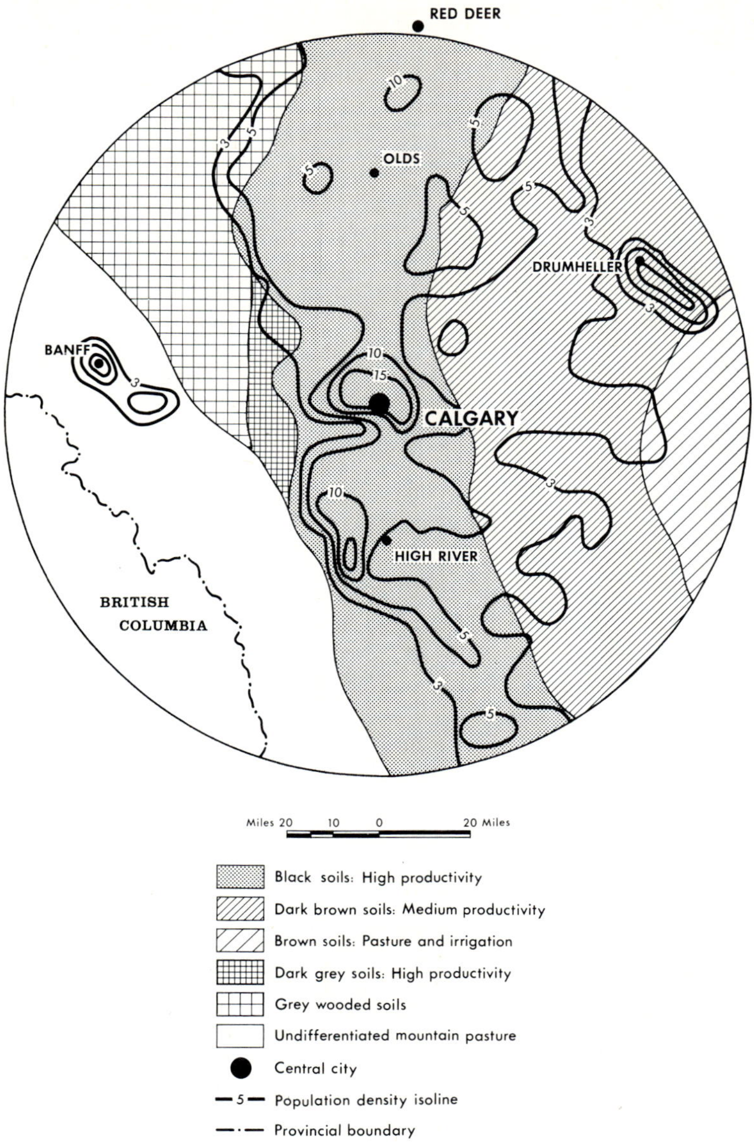

Source of soils map: *Atlas of Alberta*, University of Alberta Press, 1969.

FIGURE 2,4 Rural population density (no. per square mile) and productivity in the hinterland of Calgary, 1951.

Foothill-Prairie junction. To the west densities decline with the predominantly ranching economy of the Foothills. To the east the increasing aridity of the Prairies produces a less abrupt, but still significant decline in the agricultrual potential of the area, and with it lower population densities. However, in both areas minor nodes of population concentration can be identified: an eastern one associated with the former coal mining centre of Drumheller; a western one with the coal town of Canmore and with the resort centre of Banff. Most of the settlements in the area are primarily supported by commercial activities but the presence of natural gas processing plants provide additional economic support. This boost to the local economies is particularly important near Okotoks and Carstairs; the original area of discovery in the Turner Valley-Longview zone having declined to virtual insignificance for settlement support.

All data used in the study were obtained from the 1951 and 1971 Dun and Bradstreet Business Directories. Changes in definition, as well as the character of the commercial types, or functions, between the two years make it impossible to derive an exactly equivalent set of functions for the two years. Also very specialised yet randomly distributed single establishment types occurred in the area producing the possibility of a data matrix with many zeros. This problem was resolved by only including those commercial types that had more than two outlets in the study area in the two years. This resulted in the identification of sixty-two establishment types (or functions) in 1951 compared to sixty-eight in 1971.

As a preliminary measure of the commercial importance of the places, Functional Indices (F.I.) for each settlement were calculated for both years according to the method described by Davies:[12]

$$F.I. = \frac{\sum_{i=1}^{J} (C_i)}{J}$$

Where $C_i = \frac{t_i}{T_i} \cdot 100$

C_i =Centrality value of function i

t_i =Number of establishments of type i in settlement

T_i =Total number of establishments of type i in region

J =Total number of different establishments for each period.

These indices, although crude, provide a summary measure of the centrality, or the relative importance of each place, and because they are adjusted by the number of establishment types, they represent a percentage value comparable between the years.

The presence of groups in the matrix of establishments and places was identified by a hierarchical clustering routine.[13] However, all the clustering methods demand that the matrix of raw data is first converted into a symmetrical similarity matrix in which each place is compared to every other place. In this case a measure of Euclidean Distance provided the similarity measure. From among the range of possible clustering methods Aitcheson[14] and Sokal and Sneath[15] have observed that one of the most satisfactory approaches in Ward's method[16] of hierarchical clustering. This method, which is a minimum variance procedure, was applied to the matrix of similarity coefficients. At each stage an Error Sum of Squares (ESS) was derived which reflected the loss of information, or fusion summary, that results from treating the observations, in this case places, as groups. This measure is defined as:

$$ESS = \sum_{i=1}^{n} x_1^2 - \frac{1}{n}\left[\sum_{i=1}^{n} x_1\right]^2$$

Where X_i = value of ith object of score

n = number of observations

Figures 3,4 and 4,4 show the linkage tree results of the clustering procedures. The advantage of Ward's method is that at each level of clustering the technique produces groups that display optimal minimum variance, thereby avoiding the chaining effect so common in average distance or nearest neighbour clustering methods. The disadvantage of the technique is the same as all other hierarchical clustering routines, namely that there is no statistical procedure for choosing one set of groups over another, since each grouping stage provides the optimum minimum variance resolution. Hence the researcher has to choose groups appropriate to his problem. Fortunately there is a way out of the impasse of subjectivity. Most researchers have discovered that if the loss of information score provided at each grouping stage is plotted against the number of groups a major break in the graph is usually identified. This procedure represents the best way of picking out groups. The insets in Figure 3,4 and 4,4 show that a clear break of slope can be identified for 1951 and 1971, providing five groups in both cases.

As a test upon the utility of the solution a Stepwise Multiple Discriminant Analysis was applied to the groups, and the significance of the resulting discriminant functions was established by the F ratio tests shown in Table 1,4. These show that the cluster groups had significant discrimination at the 99 per cent level for 1951, and also for 1971 if three places (Arrowood, Beiseker and Turner Valley) were re-classified as Group II rather than Group I centres in 1971. Given the number of places involved in the grouping these tests established the overall utility of the five cluster solution. But we must recognise that this can only

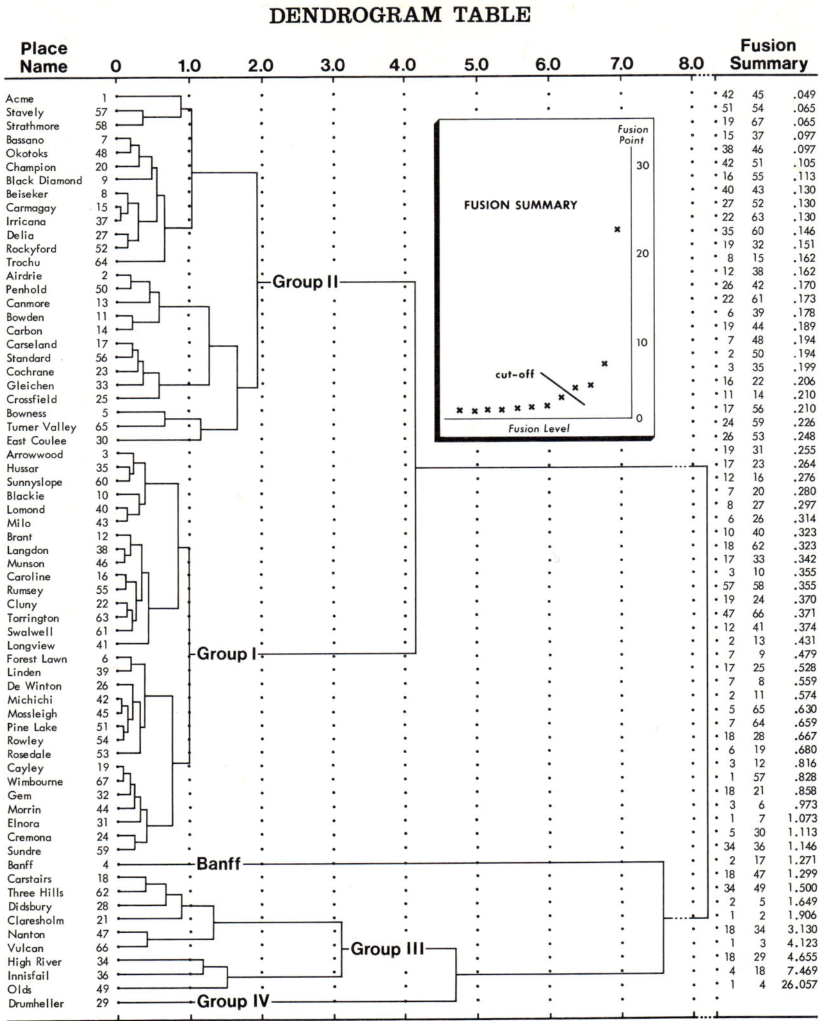

FIGURE 3,4 Dendrogram of clusters of settlements Calgary region, 1951.

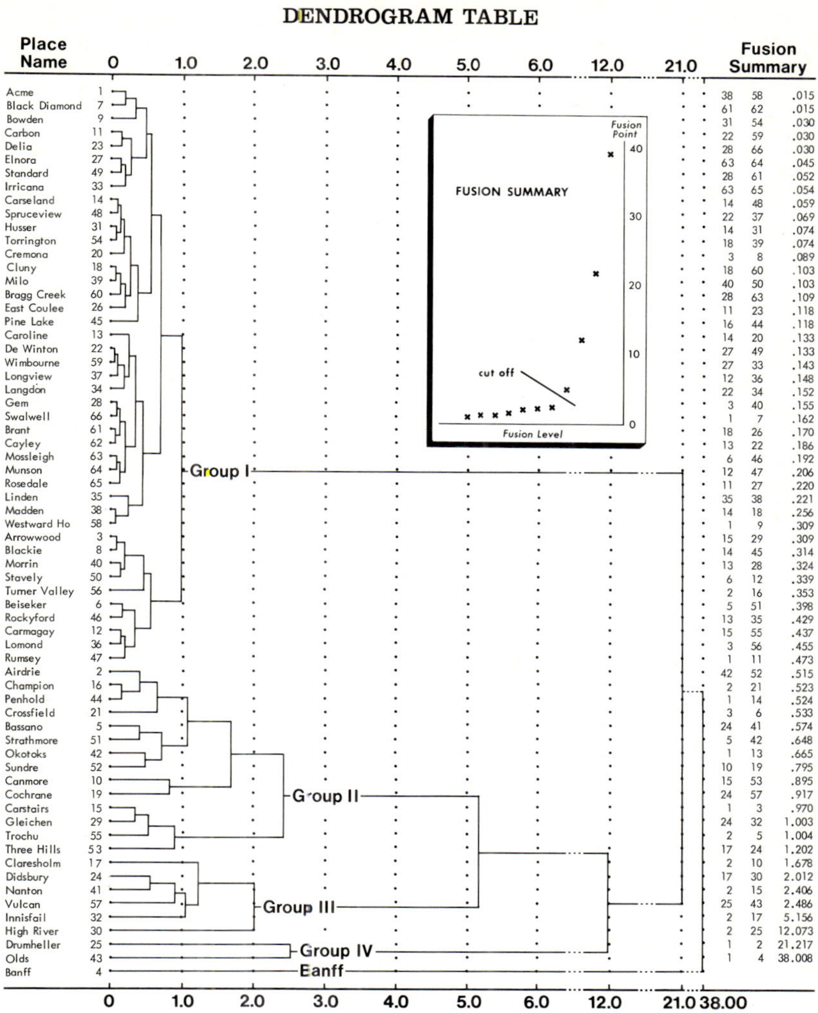

FIGURE 4,4 Dendrogram of clusters of settlements Calgary region, 1971.

be regarded as a partial test. The presence of so many zeros in the raw data matrix from higher order establishments meant that not all the commercial types could be used to calculate the multiple discriminant function. So only those commercial types present at least once in any two of the groups being compared could be used, namely thirty-one in 1951 and thirty in 1971. As only half of the establishments were used in the discriminant test comparisons, and there was only a minor revision for one of the years, it was decided to use the results as they appeared in the original cluster groupings rather than in the discriminant tests.

TABLE 1,4

F RATIOS AFTER MULTIPLE DISCRIMINANT TESTS

CALGARY 1971					CALGARY 1951				
Group	IV	III	II	I		IV	III	II	I
IV					IV				
III	6.7				III	9.9			
II	9.6	6.7			II	11.2	8.8		
I	11.4	9.3	3.4		I	18.5	25.6	10.9	

Tabulated F Ratios: 95 percent = 1.84; 99 percent = 2.39

FUNCTIONAL COMPOSITION OF THE CLUSTER GROUPS

The previous section confirmed that five clusters could be found in the data set for 1951 and 1971 but that one of these groups only contained one settlement, Banff, clearly a specialized centre. The identification of a series of clusters does not answer the question of whether the clusters have a hierarchical arrangement of functions as predicted by central place theory, or whether the clusters show distinctive specialisms according to the effect of differential spread mechanisms in a growth pole context. To deal with this question the average number of establishments present in each of the clusters was calculated, along with the frequency of occurrence of the commercial types. Establishment types, or functions, were considered to be distinctive members of a group only where the average value was over 1.0 and the frequency of occurrence was over 50 per cent. Out of the sixty-two functions analysed in 1951 only eleven were not found to be distinctive members of clusters; the figure for 1971 was fifteen out of sixty-eight.

Scrutiny of the resulting tables showed that the functions associated with each group could be regarded as being hierarchical. Not only were there a set of distinctive functions at each level, but those establishments characteristics of the smaller, or lower order place were also found at a higher level with increased frequency and with higher averages. Table 2,4 uses four functions from 1971 to demonstrate the patterns involved. For example, the Group IV settlements have all the lower order functions present, albeit with greater frequency, plus a new type, Furniture and Floor coverings, that is representative of the marginal hierarchical goods for the level. Indeed, in the complete set of functional types,[17] this progression of higher averages and more frequent occurrence is only broken by one type, General Stores. Here individual specialized commercial outlets replace certain lines of goods that are sold in the General Stores, making redundant the persistent need for more outlets of the type at higher levels.

TABLE 2,4

AN EXAMPLE OF THE HIERARCHICAL ARRANGEMENT OF FUNCTIONS: CALGARY 1971

Representative establishment types	GROUPS Mean(Freq.)*	Mean(Freq.)	Mean(Freq.)	Mean(Freq.)
Furniture floor coverings	0.05 (04.7)	0.0 (0.0)	0.5 (33.3)	1.5 (100.0)
Drug Stores	0.21 (18.6)	0.50 (50.0)	2.00 (83.3)	2.5 (100.0)
Hotels	0.77 (62.8)	1.00 (71.4)	1.17 (83.3)	4.0 (100.0)
General Stores	1.4 (81.4)	1.00 (85.7)	1.33 (66.7)	3.5 (100.0)
	I	II	III	IV

* Mean for the group, and percentage frequency of occurrence.

Figure 5,4 shows the re-arrangement of the functions in the four groups during the 1951 - 71 period. Despite changes in the nature of the establishment types, almost three quarters of the functions are present in both years allowing direct comparison between the years. But only nineteen of the commercial establishments remain at the same group level over the period. Since eleven of these are associated with Group IV it is apparent that Groups I, II, and III have a very small core of identical functions. Of the eighteen establishments that changed groups, sixteen showed an upward movement from lower to higher groups. This confirms the relevance of the upward shift of functions associated with the increasing specialisation of places described by previous workers, such as

Berry,[18] Stabler and Wil iams,[19] that is linked with the twin impact of increasing threshold size of commercial outlets and increasing consumer mobility. Only Petroleum Products and Retail Bakeries moved down in group, from Group III to II and Group IV to III respectively.

The result of the changes upon the functional composition of each group are shown in Figure 5,4. Group I, by reason of the upward shift of Car Sales, Eating Places, and Farm Equipment Sales, is reduced to the bare minimum service function of a Gas Station and General Store. The group is labelled Minimum Convenience Centers. This does not mean, however, that these places only possess these two functions; other types can be present, but they are not typical of the group as defined by the requirements of average numbers and frequency of occurrence.

The lower order functions present in Group I are found in Group II centres with greater frequency, but it is the new marginal hierarchical establishments that identify the group. In 1951 there were fifteen of these types, but these had declined to six in 1971, with only Hardware, Hotels and Groceries beirg present in both years. The presence of Hotels may appear to be an anomaly compared to other study areas. It is a consequence of the fact that licences for alcoholic beverages have been difficult to obtain in Alberta and have been rarely given to establishments that did not have hotel accommodation. The relaxation of these standards in the last five years has meant that restaurants and cafes can now obtain such facilities. The result has been to preserve the hotel function in many smaller centres. Overall, the decline in functional complexity suffered by the settlements in the Group II or Local Center cluster in the 1951-71 period is glaringly obvious from Figure 5,4.

No less than eight of the 1951 functions move up to the Group III level. In 1971 Group III had nineteen new or marginal hierarchical establishments together with the eight establishment types characteristic of Groups I and II. Any review of the range of functions present in

(Places)	Group	Functions	1971	
(43)	I	2	Gas Stations	1.3
			General Stores	1.1
(14)	II	6	Farm Equipment	1.2
			Eating Places	1.1
			Hardware	1.0
			Hotels	1.0
			Groceries	1.0
			Petroleum Products	1.9
(6)	III	19	Car Sales	3.2
			Meat Markets	1.3
			Lumber Yards	1.7
			Auto Repairs	1.5
			Drug Stores	2.0
			Male Clothing	2.0
			Electrical Repairs	1.0
			Banks	1.3
			Repairs (n.e.c.)	1.0
			Contractors	1.2
			Tire Repairs	1.0
			Jewellery	1.0
			Auto Body Repair	1.0
			Plumbing, Printing	1.7
			Heavy Construction	1.3
			Car Accessories	1.2
			Women's Wear	1.2
			Retail Bakery	1.2
			Radio, TV Repair	1.0
(2)	IV	24	Household Appliances	2.5
			Shoe Stores	1.5
			Other Retail	1.0
			Funerals	1.0
			Laundromats	1.5
			General Building	1.0
			Highway, Heavy Contruction	1.0
			Wholesale Grocery	1.0
			Wholesale Speciality	1.0
			Amusement, Sporting Goods	1.0
			Watches, Jewellery	1.0
			Furniture, Floors	1.5
			Candy	1.0
			Family Clothing	1.0
			Sports, Bicycles	2.5
			Tires, Tubes	1.0
			Construction	1.5
			Other Contractors	2.0
			Other Auto Repairs	1.0
			Wholesale Tires	1.0
			Radio and TV	1.5
			Farms, Gardens	1.0
			Plumbing, Printing	1.0
			Stationers, Books	1.0

1951		Functions	Group	(Places)
Gas Stations	1.4	5	I	(56)
General Stores	1.7			
Farm Equipment	1.1			
Eating Places	1.6			
Car Sales	1.6			
Special Contractors	1.5	15	II	(6)
Bowling, Billiards	1.3			
Hardware	2.0			
Hotels	1.2			
Groceries	3.3			
Meat Markets	2.2			
Lumber Yards	1.8			
Auto Repairs	2.0			
Drug Stores	1.7			
Male Clothing	1.2			
Electrical Repairs	1.2			
Banks	1.0			
Repairs (n.e.c.)	1.0			
Household Appliances	1.0			
Shoe Repairs, Cleaning	1.0			
Petroleum Products	1.7	8	III	(3)
Tire Repairs	1.0			
Jewellery	1.3			
Auto Body Repair	1.0			
Blacksmith	1.7			
Other Retail	1.7			
Funerals	1.0			
Cleaning, Dyeing	1.0			
Retail Bakery	2.0	21	IV	(1)
Power Laundries	1.0			
Children's Wear	1.0			
General Building	1.0			
Highway, Heavy Construction	1.0			
Wholesale Grocery	1.0			
Wholesale Speciality	1.0			
Amusement, Sporting Goods	1.0			
Watches, Jewellery	1.0			
Furniture	2.0			
Candy	1.0			
Family Clothing	2.0			
Sporting Goods	1.0			
Tires, Tubes	1.0			
Photographers	2.0			
Other Contractors	2.0			
Groceries with Meat	5.0			
Machine Shops	1.0			
Camps	1.0			
Metal Dealers	1.0			
Milk Retailers	1.0			

N.B. Figures after the function show the average number for that group or cluster.

↖ Downward Shift to Lower Grade in 1971

↙ Upward Shift to Higher Grade in 1971

↔ No Change

FIGURE 5,4 Functional composition of four clusters, 1951-1971.

Group III indicates a variety of types capable of servicing weekly or two weekly shopping needs, sc they are labelled Shopping Towns. However, the changes between 1951 and 1971 has been quite abrupt. Only three types, Tire Repairs, Jeweller, and Auto Body Repairs, are found in both years. It is the upward shift of nine lower order functions from Groups I and II in 1951 that give the group its variety. The result can only be described as a 'squeeze' on these middle ranking places with the variety of Shopping Goods obtainable in Group II in 1951, now being found in Group III places in 1971. This leaves Group II places as local centres providing specialized convenience level functions in 1971, whilst the Group III places in 1951 were really a specialized shopping or specialized Group II grade.

Given the small number of towns in the Group IV category there is little point in repeating the details of the centres shown in Figure 5, 4. It is enough to observe that both Group IV places have over twenty new establishment types, with eleven functions present as a core in both years, including wholesale establishments for the first time. Banff, the sole member of the fifth group, is shown in the analysis as a separate specialized group in both years. Table 3,4 lists a representative range of the variety of establishments found in Banff. It emphasises the range of specialization in the resort, ski and recreation centre by showing that it contains over eighty per cent of the camera and gift shops, and over twenty per cent of the eating places, hotel and tourist accommodation listed in Dun and Bradstreet outside Calgary.

THE GROUPING OF THE PLACES

Figure 6,4 shows the major changes that have occurred in the composition of the four groups according to the places, rather than the functions, as well as the percentage Functional Index of all the places in Group II and above in 1971. In total these indices confirm, yet

TABLE 3,4

EXAMPLES OF THE SPECIALIZATION OF FUNCTIONS IN BANFF, 1971

Establishment Type (Functions)	Numbers In			
	Banff	Group IV Average	Whole Region	Percentage in Banff
Eating Places	25	10	125	20.0
Hotels, Trailer Courts	20	4	82	24.4
Gift, Souvenir Shops	13	0.5	15	86.7
Gasoline Service Stations	11	10	154	6.5
Sports, Amusement	6	1.0	14	42.9
Camera, photo graphic equip.	5	0.5	6	83.3
Drug Stores	4	2.5	37	10.8
Photographic Studios	3	0.5	7	42.5
Plumbing, paint-ting masonry	3	5.0	41	7.3
Automobile Rental	2	0	2	100

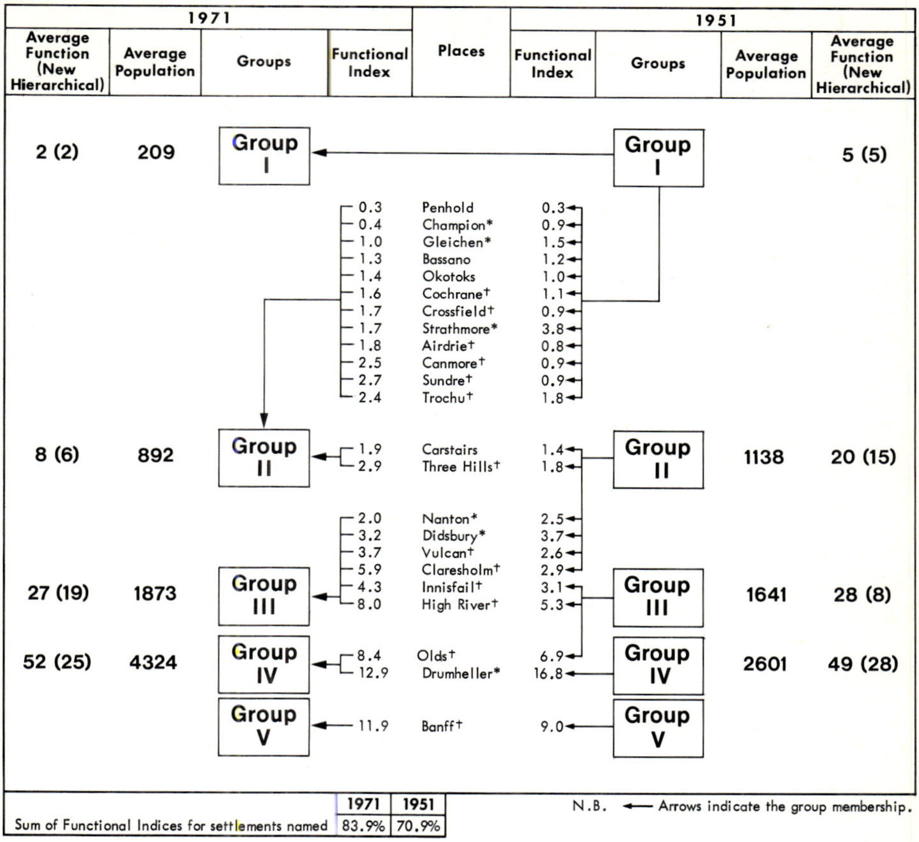

Changes in Index Status >± 0.5 are shown as follows: †Gain; *Loss.

FIGURE 6,4 Changes in status of the largest places.

again, the increasing centralisation that has taken place in the central place system, for the largest places account for 83.9 per cent of the centrality of the functions in the area in 1971 compared with only 70.9 per cent in 1951. Of the fifty six Group I centres present in 1951 with an average population of 209, twelve places move out of the group to the next level, leaving forty three places in the lowest size group in 1971. The average population size for the group in 1971 is 209, a figure almost identical to that obtained by Stabler and Williams[20] in their study of the Qu'Appelle Valley in Saskatchewan in 1970.

Parallelling the decrease in the functional complexity of the Group II places over the period 1951-1971 is the general drop in the average population level of its members from 1138 to 892, although the upward movement of the Group I places means that the size of the group increased from six to fourteen centres. Only Carstairs and Three Hills remain as members of the group for both years, with Vulcan, Nanton, Didsbury, and Claresholm moving up to the next grade of centre.

The upward movement of functions from Group II to Group III has already been described, and represents the consolidation and strengthening of these Group III places as centres with a wide range of shopping types. Not surprisingly this is associated with a rise in the average population level of the centres from 1641 to 1873, the latter figure again parallelling the findings of Williams and Stabler[21]. Only Innisfail and High River remain as consistent members of the group, and Olds moves out of the group to join Drumheller as the second Group IV centre in the region.

In spatial terms, the primary result of these centralisation tendencies has been to increase the average distance apart of the Group I places from nine to fourteen miles. For the other higher order groups, average distance calculations are rather meaningless; Figure 7,4 shows that they are not evenly spread throughout the region, but are concen-

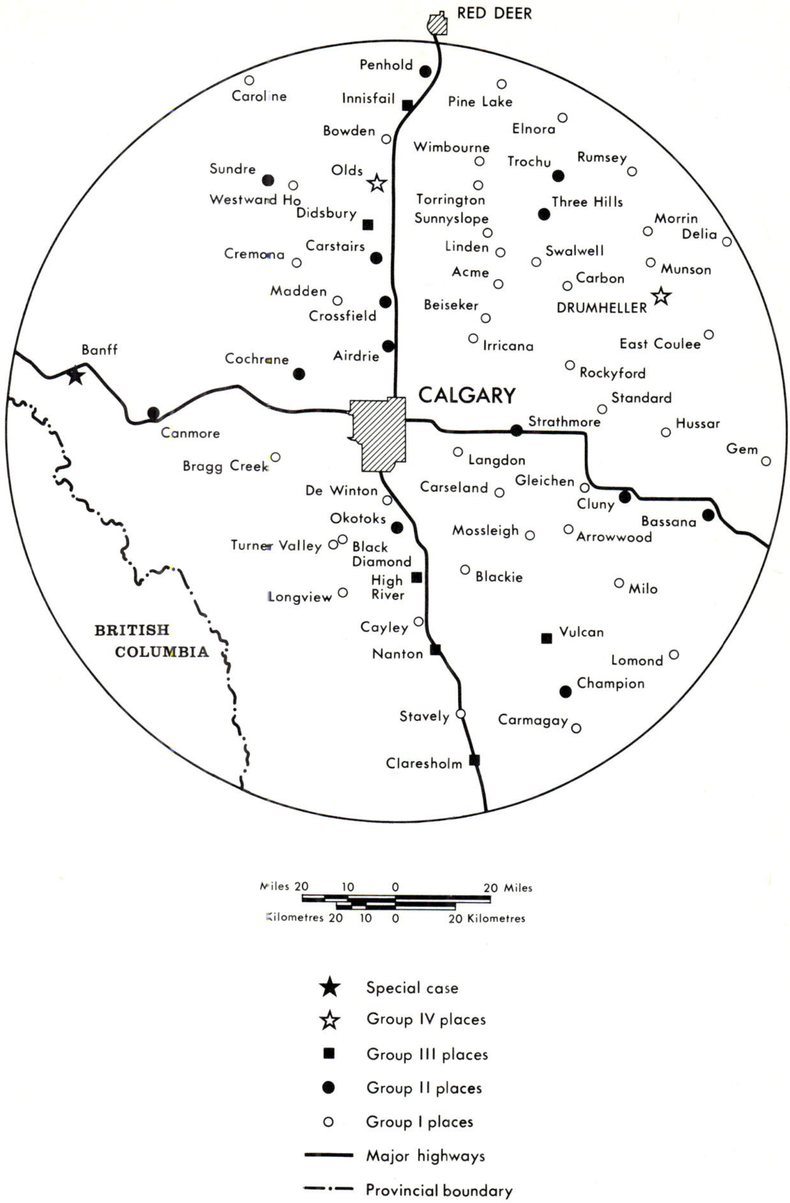

FIGURE 7,4 Distribution of settlements by groups Calgary region, 1971.

trated along the main north–south highway running through Calgary, and to a lesser extent along the east-west Trans Canada Highway. Only Three Hills, Trochu, Vulcan, and the former coal town of Drumheller, in the interstices of the four sectors, form exceptions to this trend. The result is to produce a pattern rather similar to Lösch's 'city rich-city poor' sectors around the metropolitan centre. Yet however similar the pattern, it is unlikely that the explanation for this Calgary example corresponds to Lösch's theoretical rationale for 'city rich' sectors, since this was the result of theoretical networks that are not fixed by the presence of settlements. Thus, Figure 7,4 shows that there are plenty of settlements between the major sectors; the problem is that these settlements are of comparatively low order. Three factors can be suggested as explaining the pattern. Firstly, the decline of rural densities in the ranching and forest lands of the west and the drier agricultural areas to the east must reduce thresholds, so rural centres can rarely develop without a major boost from some additional economic activity such as a mine. Secondly, the main road settlements had the advantage of an earlier commercial start over the other places, an advantage confirmed by the pattern of metalled roads in the region. Thirdly, the absence of serious rivals to the main road settlements means that consumer trade patterns continue to focus on these places. This produces the east-west orientation of trade flows demonstrated in Figures 8,4 and 9, 4 which show the usual settlement frequented by a sample of 241 farm households for Food Good and Major Applicances in 1971. Incidentally, Figure 9,4 also demonstrates the dominance of Calgary in the purchase of higher order goods, in this case in areas over 60 miles from the city, confirming the decision to use an extensive boundary to the study area.

The changes in the individual Functional Indices of the places shown in Figures 6,4 and 7,4 confirm these overall generalisations. The relative decline of Drumheller may be attributed to the economic base

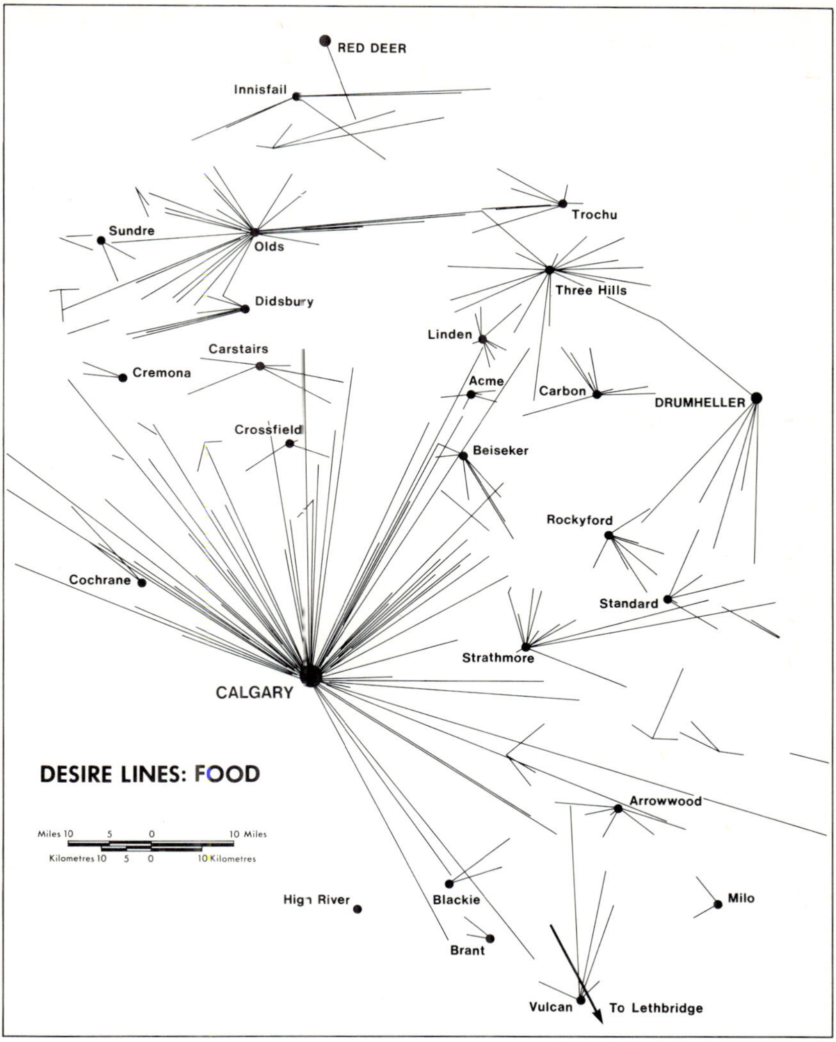

FIGURE 8,4 Shopping movements for food, 1971.

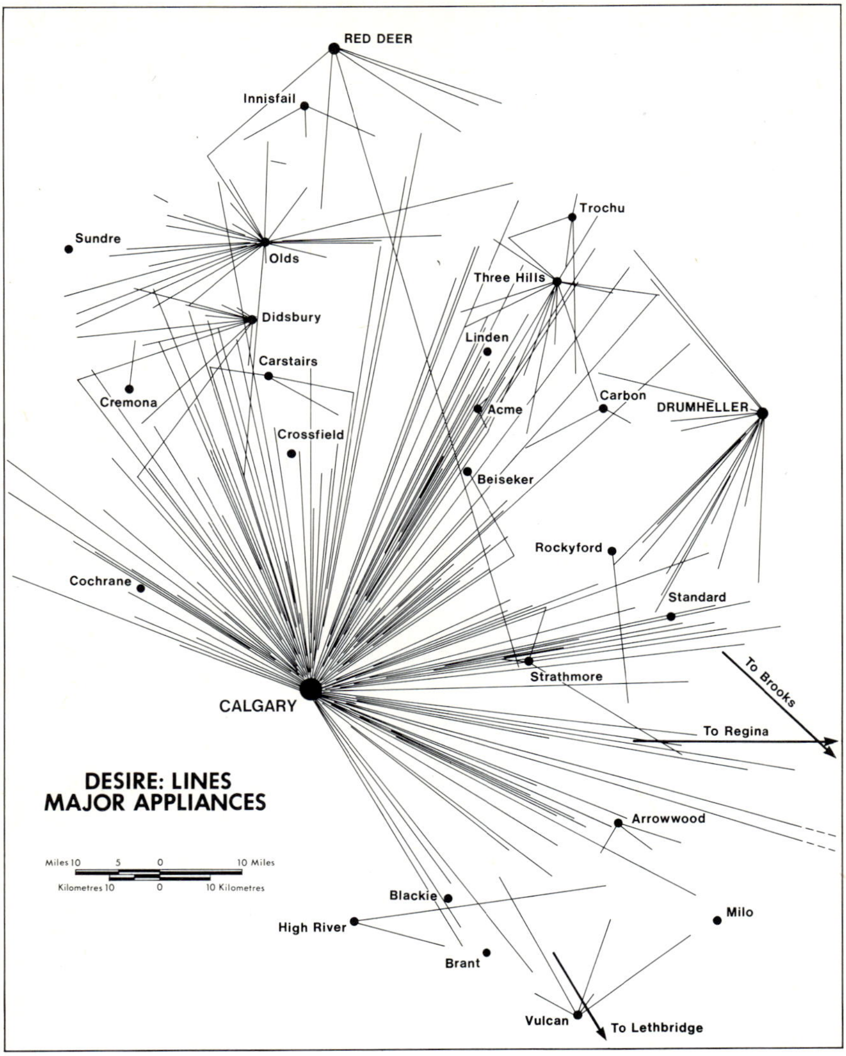

FIGURE 9,4 Shopping movements for major appliances, 1971.

PLATE 15
New building, adjacent to the City Hall, eastern C.B.D.

PLATE 16
High-rise apartments adjacent to major regional shopping centres.

City of Calgary Photo

changes consequent upon the closure of the coal mines in the area during the period. But apart from this town it is the eastern centres, for example, Champion, Gleichen and Strathmore, that have suffered the greatest losses in relative importance among the largest places. However, it must be emphasised that the decline of the smallest settlements of all has been even greater, so that these three places are still classified as Group II places in 1971. Added to this set of declining places must be the towns of Nanton and Didsbury. In both cases they have maintained their status as weekly shopping centres, with a re-classification as Group III places in 1971. However, in Functional Index terms they have declined in relative importance, probably because their trade areas have been squeezed out by competition from neighbouring higher order centres, that is High River and Claresholm in the case of Nanton, and Olds and Calgary in the case of Didsbury. Again, however, a non-commercial factor must be added to the explanation. For example, the closure of the R.C.A.F. base in Nanton in the mid 1950's had a major effect upon the economy of the town, a change not counteracted by the expansion of its trade area, or by the addition of the type of industrial activity attracted to centres such as Claresholm.

Contrasting with this pattern of decline is the fact that most other larger places have increased their relative importance, illustrating once again the centralisation tendencies within the system. Yet the growth of two centres in the Foothills-Mountain zone, Sundre and Canmore, as well as three places around Calgary, Okotoks, Airdrie and Cochrane, must also be noted. In the case of Sundre the reason lies in the normal development of a new trade centre in the Foothills, plus the boost given by the recreational growth of the area. The expansion of the coal industry is the cause of Canmore's growth. With respect to the three places around Calgary it is clear that none of them is large enough to compete even with one of the smaller community level shopping centres in the city, so their development is at the convenience good level. These results, therefore, show no sign of the growth inhibition tendencies associated with polarization mechanisms as recognized by Hodge in Saskatchewan.[22] Indeed, in the period since 1971, all three centres have added new housing estates. These are occupied, in the main, by commuters from Calgary, whilst Airdrie is to be the recipient of functions decentralized from the capital at Edmonton under the Provincial Government's Dencentralization Policy.

CONCLUSIONS

This study of the changes that have occurred in the commercial importance of settlements around Calgary over the twenty years from 1951-1971 has been a 'before and after' study. It is a description of the results of the changes, rather than a study of the processes of change. Yet despite this limited perspective it has demonstrated that a hierarchical pattern of central places is characteristic of the area during both years. Apart from the case of Banff there is no sign of the growth of a set of specialised centres of the type postulated by Lösch, or demonstrated empirically by Pahl, or of any inhibition of growth around Calgary. However, the spatial pattern of centres is shown to be very similar to a sectoral distribution, in a manner reminiscent of Lösch's 'city rich-city poor' sectors. But this pattern of concentration is based on the size of places, not on their presence or absence as in Lösch's study, and is closely linked with the underlying distributional patterns of population and road.

In general, the changes found over the period show the familiar story of the increasing centralisation of commercial functions in the larger rural service places, a feature that would be even more obvious if Calgary itself were added to the analysis. This centralisation takes the form of an upward shift of functions, with a crystallisation out of certain places because of the 'squeezing out' of the middle range of places. Group III centres replace Group II as the first group of centres with a wide range of shopping goods. From a consumer viewpoint these requirements can only be satisfied at a Group III level in 1971, whereas they were present at the Group II level in 1951. But a word of caution must be introduced in describing these changes in the usual way as an 'upward shift' of functions. This could give a false impression of the actual process of change. If two sets of places, A and B, are considered, and A has declined, it is not so much that functions moved from A to B, but that B kept functions and A lost them. So although a re-classification of places would give the appearance of an upward shift, it is the decline of the A centres that is the key to the change. So the application of growth centre ideas to changes in central place structure may not operate through the diffusion of new functions, but through the differential loss of older ones. It may be suggested that the development of a viable and integrated theory of the inter-urban system may require not only the study of process of spread or diffusion, but the process of decline, or reverse diffusion, a noticeable omission in the present literature.[23]

REFERENCES

1. All the analyses were carried out on the University of Calgary C.D.C. 6400 system using the following package programmes:

   (a) BMD 07M: Stepwise Multiple Discriminant Analysis in: DIXON, W. (ed.) BMD Biomedical Computer Programmes. Berkeley and Los Angeles :University of California, 1968;

   (b) WISHART, D "Cluster Analysis, "Kansas Geological Survey Computer Contributions, No. 38, Lawrence, Kansas, 1964.

2. CHRISTALLER, W (Translated by BASKIN, C.) Central Places in Southern Germany. Englewood Cliffs : Prentice Hall, 1966.

3. HANSEN, N.M. Growth Centres in Regional Economic Development. New York : Free Press, 1972.

4. LÖSCH, A (Translated by WOGLOM, W.H. and STOLPER, W.F.), The Economics of Location. New York : Yale University Press, 1954.

5. HODGE, G. "The Prediction of Trade Centre Viability in the Great Plains, "Papers and Proceedings of the Regional Science Association, XV (1965) pp. 87-115.

6. PAHL, R.E. "Urbs in Rure : The Metropolitan Fringe in Hertfordshire", L.S.E. Geographical Paper No. 2 (1964), London School of Economics (1964).

7. MOSELEY, M.J. "The Impact of Growth Centres in Rural Regions," Regional Studies, 7 (1973), pp. 57-94.

8. HAGGETT, P. Locational Analysis in Human Geography. London : Edward Arnold, 1965.

9. BERRY, B.J.L. Geography of Market Centres and Retail Distribution. Englewood Cliffs : Prentice-Hall, 1967.

10. STABLER, J.C. and WILLIAMS, P.R. "The Changing Structure of the Central Place Hierarchy," Land Economics, (November, 1973), pp. 455-458.

11. HODGE, G. op. cit.

12. DAVIES, W.K.D. "Centrality and the Central Place Hierarchy," Urban Studies, 4 (1967), pp. 67-79.

13. WISHART, D. op. cit.

14. AITCHESON, J.A. "The Farming Systems of Wales," Paper presented to I.G.U. Commission on Agricultrual Typology, Verona, Italy, 1974.

15. SOKAL, P. and SNEATH, R. Numerical Taxonomy. San Francisco: W.H. Freeman, 1973.

16. WARD, J.H. "Hierarchical Grouping to Optimize an Objective Function," Journal of the American Statistical Association, 58, (1963), pp. 236-244.

17. GYUSE, T.T.I. "Service Centre Change in Metropolitan Hinterlands", Unpublished Master's Dissertation, The University of Calgary, 1974.

18. BERRY, B.J.L. op.cit.

19. STABLER, J.C. and WILLIAMS, P.R. op. cit.

20. Ibid.

21. Ibid.

22. HODGE, G. op. cit.

23. The authors gratefully acknowledge the financial help provided by the University of Calgary Grants Committee Award No. 2374 towards th cost of carrying out this study.

PLATE 17
New public buildings, eastern C.B.D.

PLATE 18
View from the Bow River, looking eastwards.

City of Calgary Photo

PLATE 19
The Planetarium.

PLATE 20
Modern single family detached dwellings in a middle income community.

City of Calgary Photo

SECTION II

INTERNAL SOCIO-ECONOMIC STRUCTURE

"It is difficult ,and perhaps impossible, to supplant the role of the market in determing the spatial pattern of retail facilities in suburban areas".

"The concept of the shopping centre, implying interaction between dwellings and contiguous groups of stores was not supported by the interview data".

"...the social dimensions of the community areas of Calgary conform very closely to the models proposed for other western cities, despite the local and individual peculiarities of Calgary's growth."

PLATE 21
Glenmore Reservoir on the Elbow River.

City of Calgary Photo

CHAPTER 5

PLANNING FOR RETAIL SERVICES IN NEW RESIDENTIAL AREAS SINCE 1944

D.G. Harasym
and
P.J. Smith

University of Alberta

It has long been held by urban planners, as one of the most basic of planning principles, that the complete array of low-order convenience services must be incorporated into the design of new residential areas.[1] Planners tend to think in terms of self-contained service units, commonly referred to as communities or neighbourhoods, in which the location/allocation problems of schools, parks and playgrounds, convenience shops, churches and social service outlets receive as much attention as the design of the actual housing areas. In fact, one of the major responsibilities of the <u>city</u> planner (that is, the planner employed by civic government) is to ensure that the accessory features of residential development are adequately provided for.

In Calgary, this general conception of residential areas as community service units has prevailed for more than 25 years. In this essay, though, attention is focused on only one of its elements: the provision of local shopping facilities. The intent is to determine how planning policies have affected the form, function and spatial arrangement of retail services in the residential districts that have been developed in Calgary over the past 30 years. It must also be understood that the emphasis is on planners' <u>perceptions</u> of community service needs in new residential areas, and the consequences of their attempts to translate these perceptions into practice. In addition to providing a partial explanation of the contemporary spatial pattern of retail facilities, then, it is possible to reach some conclusions about the role of urban planners in the process of land use evolution.

DATA SOURCES AND ORGANIZATION

The prime data sources are the minutes of three committees that have had the key roles in Calgary's planning machinery: the town planning commission (1944-1951), the technical planning board (1951-1963) and the municipal planning commission (since 1963). Other sources include files in the City Clerk's Department; City Planning Department reports and files; newspaper articles; and interviews with land developers and civic officials.

From these sources it quickly became apparent that postwar residential planning policies can be classified into three distinct periods:

1. The period of the modified gridiron pattern, 1948-1953.
2. The period of the neighbourhood unit concept, 1954-1963.
3. The period of the sector concept, since 1963.

The main characteristic of the first period is that it was a time of transition, in which the legacies of an unplanned past were being overcome and the first formal expressions of urban planning philosophy were struggling into coherence. In the second period, the notion of integrating community services with the design of residential areas was both accepted and implemented. Then came a reorientation of planning philosophy that marked the break to the third period; the community service concept was retained, and even strengthened, but a new conception of areal scale was introduced, with radical consequences for the spatial form of suburban Calgary.

In view of the clear sequence of planning periods, it was only logical to follow a chronological organization for this essay. This approach is also suitable because the chronological sequence is an evolutionary one in so far as the planners' perceptions of community service needs is concerned.

THE PERIOD OF THE MODIFIED GRID PATTERN, 1948-1953

With one or two exceptions, most notably in the prestigious suburb of Mount Royal, land subdivision during Calgary's first 75 years was based on the grid system. Moreover, most of the subdivisions were registered during the frantic land boom years of 1908 to 1912.[2] When the optimism evaporated, the vacant subdivisions stretched for miles into the open countryside. Sometimes the streets were made up, and utility lines and streetcar tracks were laid, but mostly the face of the countryside remained untouched: the subdivisions were paper ones, and the land titles were held by absentee speculators.

During the erratic but generally slow growth of the next 35 years, the land boom subdivisions provided all the building sites that were needed. Most, in fact, were still vacant in 1948, when the new but more prolonged boom of the oil years began to crest. Like other western Canadian cities at the time, the outer edge of Calgary in 1948 was frayed out through a broad and largely sub-standard periurban zone, beset with all the usual problems. Fortunately, their lesson was not lost. With the dawning realization that rapid urban growth was once again inevitable, more critical attitudes towards development began to take shape, along with a determination that the mistakes of the past would not be repeated.[3] These attitudes took root at two levels. In the first place, the provincial government reacted very quickly to try to provide more effective mechanisms of urban land use planning and development control. Planning legislation had existed in Alberta since 1913 but it had had no force. Then, through amendments in 1948 and 1950, the government began to give warning of a tougher spirit, and signposted the way to a complete reworking of its planning legislation, in the Town and Rural Planning Act of 1953.

Meanwhile, comparable concern was being demonstrated within Calgary. The town planning commission, an advisory body that was set up in 1929, became a more active proponent of planning, particularly after 1944.[4] An intermittent press campaign also brought some pressure on city council.[5] A town planning engineer (actually a land surveyor) was appointed in 1946; a planning consultant was engaged in 1949 to advise on planning structures for Calgary;[6] and a full-scale planning department was organized in 1950. Then, in 1951, city council took advantage of the 1950 amendment to the planning act to replace the old advisory town planning commission with a technical planning board which was to become the major decision-making body on planning matters, particularly with respect to the design and development of new residential areas.[7] Initially, it too was advisory but its composition (heads of those civic departments most directly concerned with physical development) almost guaranteed that it would assume much greater authority than its predecessor. This was assured some months later. Under the provisions of an interim development order,[8] city council gave broad discretionary powers to the city planning department and the technical planning board to rule on applications for land subdivision and development.[9] These were intended to be "interim" powers, while the first general plan was being prepared, but they were to remain in force for many years.

Paralleling these procedural and structural changes, the outlines of a residential land use planning philosophy also began to take form.

As early as 1945, the city commissioner urged the town planning commission to give thought to ways of ensuring that future residential areas should be adequately provided with schools, playgrounds and commercial sites.[10] By 1947, the Calgary Public School Board was beginning to buy school sites in advance of housing construction, to make certain that new schools could be built in the most convenient situations.[11] In 1950, there was a great deal of discussion about neighbourhood park needs and standards.[12] Central Mortgage and Housing Corporation was also an important influence after 1948; through its mandate to foster "sound community planning," C. M. H. C. was the first active promoter of what are known today as planned unit developments.[13] From a number of different directions, then, and for a variety of motives, the notion of the residential district as a community service unit was beginning to emerge.

These piecemeal, largely intuitive gestures were pulled sharply together in 1951, in one of the first reports to be written by the new planning director.[14] In the name of "orderly growth" he offered the opinion that the neighbourhood unit concept should be adopted as the basis for the spatial and functional organization of suburban development. In this, he was borrowing directly from the work of Clarence Perry for the New York Regional Plan in the 1920s.[15] Although the principles of neighbourhood convenience service had been known for very much longer, it was Perry who converted them into the clear philosophy of residential design that is summarized in his well-known diagram of the model neighbourhood unit (Figure 1,5). It was this that the Calgary planning director had in mind, as the following statement makes plain:[16]

> Such a (residential) unit would be the area reasonably served by the average elementary school. It is hoped that such units can be developed so that a neighbourhood spirit is engendered through common interest in

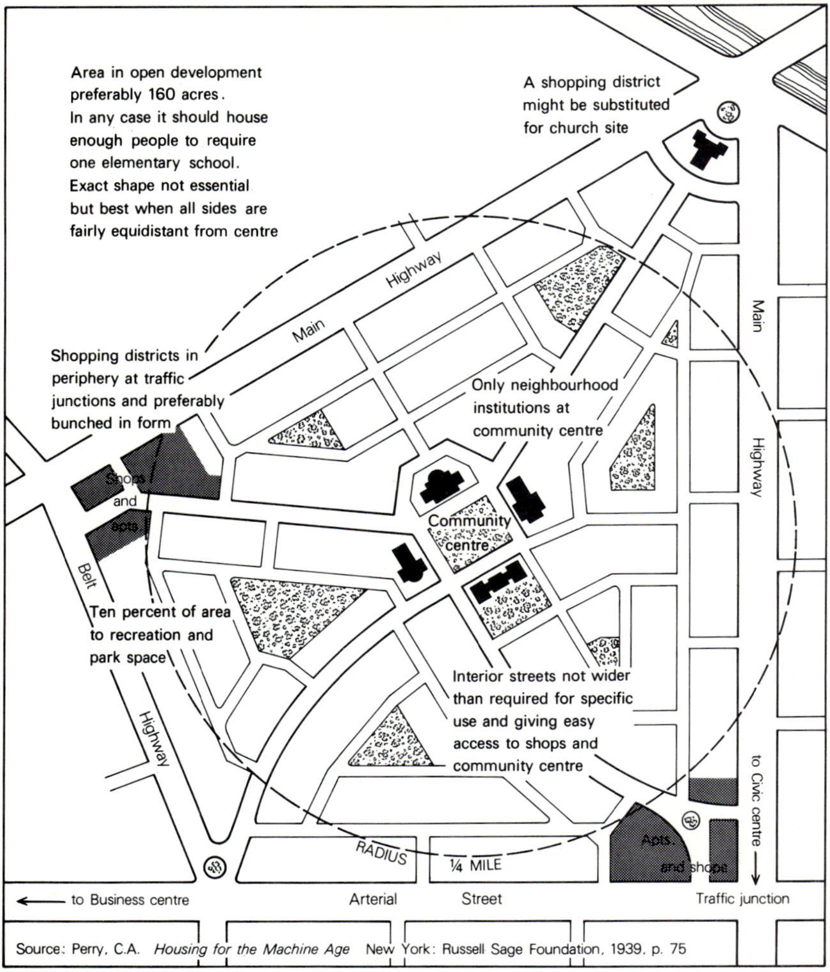

FIGURE 1,5 Neighbourhood unit principles.

> immediate local affairs, that they will be areas that are not to be invaded by major traffic streams and will therefore produce the safe and healthy living environment set out as an objective . . . It is also hoped that these unit areas will have ready access to good thoroughfares and transit service and that good local shopping will be quite readily available. These are understandably contributing features in making an area a thoroughly desirable place to live.

It was pure Perry, and it seems to have struck an immediate receptive chord in civic officials and land developers alike. Its implementation, though, was hindered by the persistence of large tracts of partially developed land, which absorbed the early post-war demand for new housing. Although it was possible to make some concessions to neighbourhood unit principles, and even to modify the grid pattern of subdivision to some extent, the result was an obvious compromise. Some attention was at last being given to the allocation of community service sites in relation to the residential areas that they were to serve, but it was not yet possible to create communities with clear physical identities.

Shopping Facilities: the Knob Hill Case

The extent of modified grid development, in which there was at least partial use of neighbourhood unit principles, is shown in Figure 2,5. A detailed example, the Knob Hill district of southwestern Calgary, is illustrated in Figure 3,5. From this it can be seen that although the street network has obviously been altered in a number of critical ways, the lines of the original subdivision are by no means lost. Neither has it been possible to mould a distinctive physical unit from the undifferentiated grid matrix.

Another general point is that the opportunity was taken to vary the lot pattern as well as the street pattern. Instead of assuming that

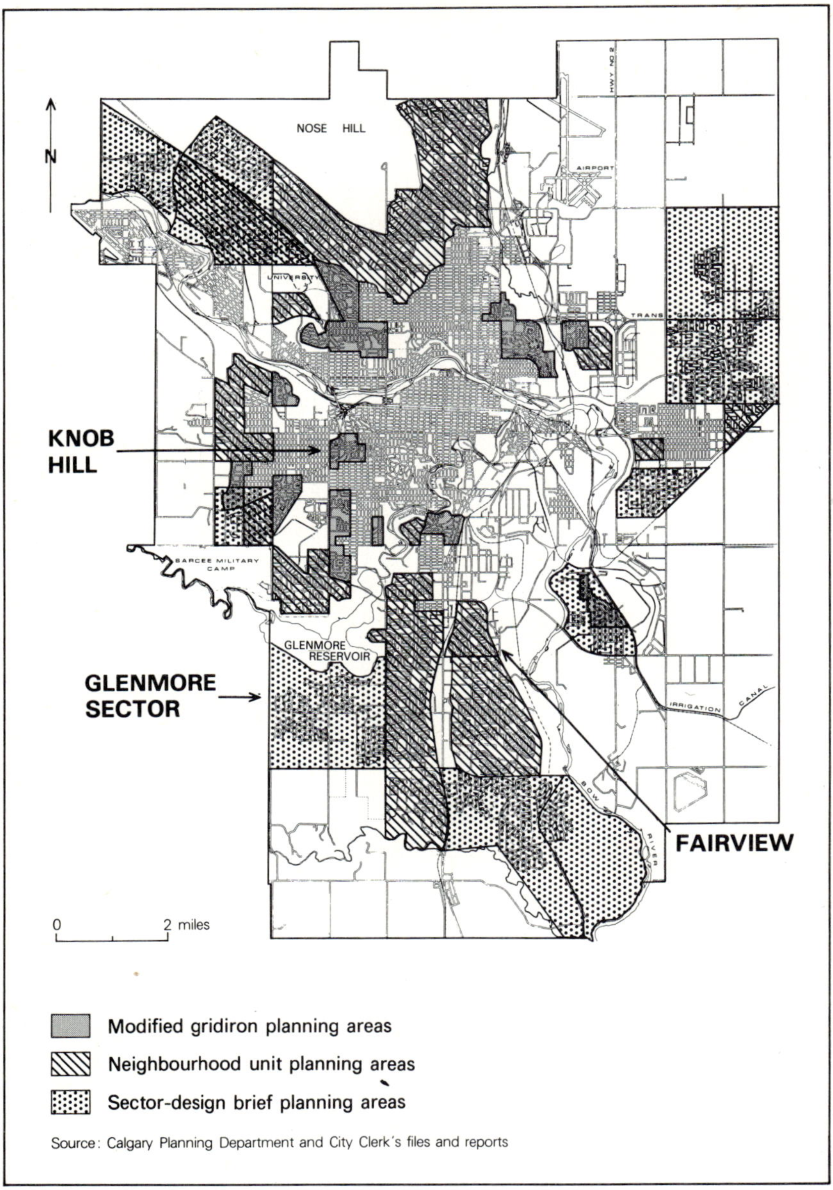

FIGURE 2,5 Modified gridiron neighbourhoods, neighbourhood units, and approved sector/design brief planning areas, 1948-1974.

grocery store, a drug store and a barber's shop. By 1953, though, the site on Richmond Road was still unoccupied; it was therefore rezoned for residential use.[17] Then, at some later stage, the three shops that were grouped together between 22nd and 23rd Avenues were converted to a soft-drink outlet, an organ sales centre and a real estate office. One of the service stations also changed function, to a tire sales depot. In other words, the local convenience function of the Knob Hill facilities gave way to services that were oriented to much larger catchments than the single neighbourhood.

These changes, though miniscule in themselves, were symptomatic of inadequacies in the planners' perceptions of retail space needs. On the one hand, they underestimated the caution of the entrepreneurs; on the other, they were unable to anticipate the retailing revolution that was about to burst on Calgary, in the era of the supermarket and the planned shopping centre. When the commercial zoning for Knob Hill was proposed by the town planning engineer to the town planning commission in 1949,[18] retailing in Calgary was still overwhelmingly concentrated in the central business district. There was some ribbon development along the main access streets and at the termini of streetcar tracks, but otherwise there was only a scattering of isolated corner stores. The notion of constructing convenience shops in phase with residential construction in suburban locations was simply not in the minds of the entrepreneurs at the time. Through the 1950s, in fact, they consistently failed to take advantage of the permissive zoning of suburban sites, preferring to cling to their conventional locations in and near the city centre. The rezoning and residential development of the Richmond Road site was just one small localized consequence.

At the same time, the planners were also constrained by convention. The Knob Hill commercial sites, for example, presumed to do no more than to group the conventional neighbourhood services in a conventional street-front layout. The sites were certainly not adequate for shopping centre development, and the small shops that were built on them could not compete with the shopping centres that were opened nearby a few years later. Hence their change of function to higher-order services that were attracted mainly to a location on a busy arterial street.

THE PERIOD OF THE NEIGHBOURHOOD UNIT CONCEPT, 1954 - 1963

Most of the immediate post-war growth in Calgary was absorbed in the partially developed subdivisions or in modified grid neighbourhoods like Knob Hill. By 1952, though, the serviced land supply was nearly exhausted, and the infilling process had largely run its course. There was still some periurban outliers, but the main development thrust was about to be concentrated on virgin land, most of which had known no permanent economic use in more than forty years. Land boom subdivision had been followed by tax delinquency in the years after World War I. Large tracts of land were forfeited to the City of Calgary which then, in the 1950s, was able to assume almost absolute control over their development. Of particular relevance to this essay, it had virtually free hand to cancel the paper subdivisions and replace them with newly-designed layouts on neighbourhood unit principles.

By the time this threshold was reached, Calgary's planning machinery was well established.[19] The neighbourhood unit had also come to full respectability, as the basic concept behind the functional and spatial organization of new residential development. Quite apart from the influence of the planning director in the technical planning board, there was pressure from external sources. Central Mortgage and Housing Corporation, because of its financial interest in most of the new houses that were built, insisted on the right to review all residential subdivision plans. Design modifications were made frequently to meet C.M.H.C. requirements or objections.[20] Although C.M.H.C. never insisted that new neighbourhoods conform exactly to Perry's physical model (Figure 1,5), Perry's principles were clearly endorsed. The same could be said of the provincial government's subdivision regulations which were adopted by order-in-council in 1953.[21] Though mostly devoted to such mechanical details as minimum road widths and lot sizes, they also incorporated a number of notions that were fundamental to the neighbourhood unit concept. These included the compulsory dedication of land for community reserve purposes (schools and parks); maximum safety and privacy in the design of the street system; and convenient relationships among the various neighbourhood elements.

The subdivision regulations also established an important planning procedure: the technical planning board was made the subdivision approval authority, so that the residential planning process came very close to being self-contained within the civic administration. The land supply was effectively a municipal monopoly; the phasing of land development was under the control of the civic utilities departments; and the neighbourhoods were designed by the planning department for approval by the technical planning board. Some other governmental agencies, such as the two Calgary school boards, also had access to the approval process, but inputs from private industry were almost non-existent. This situation changed steadily through the 1950's, as the municipal land reserve was depleted and attention shifted to privately-owned land under the control of development companies. Even so, the convention of the city planning department's involvement in residential design was maintained. It was not just that the entrepreneurs made no serious attempt to depart from neighbourhood unit principles; they relied on the planning department to provide them with schematic outlines of land use and street patterns in new residential areas, and even to fill in much of the design detail.[22]

Shopping Facilities: The Fairview Case

An example of a planned neighbourhood unit is displayed in Figure 4,5. This is Fairview in southeastern Calgary, a neighbourhood which was designed and developed in the late 1950s. It was actually one of the first privately-planned neighbourhoods, but its planner had previously been the City's subdivision designer and so was fully familiar with the procedures and expectations of the technical planning board. With respect to retail facilities, Fairview certainly exemplifies the planning perceptions of the time, as well as the difficulties of implementing a particular planning concept.

By the time that the Fairview neighbourhood unit was planned, the notion of the unified shopping centre had come into common currency, though the planners' conception of the scale of the neighbourhood shopping centre was still quite imprecise. The Fairview shopping centre, as developed, contains twelve separate outlets with a clear emphasis on low-order retail, personal and recreational services (for example a small food market, two hair stylists and a bowling alley). In the original plan, though, a substantially larger centre had been envisaged, on a site roughly twice the size of the present one.[23] This was by no means an uncommon experience, and the minutes of the technical planning board contain frequent references to applications for the rezoning of portions of shopping centre sites, usually to permit apartment construc-

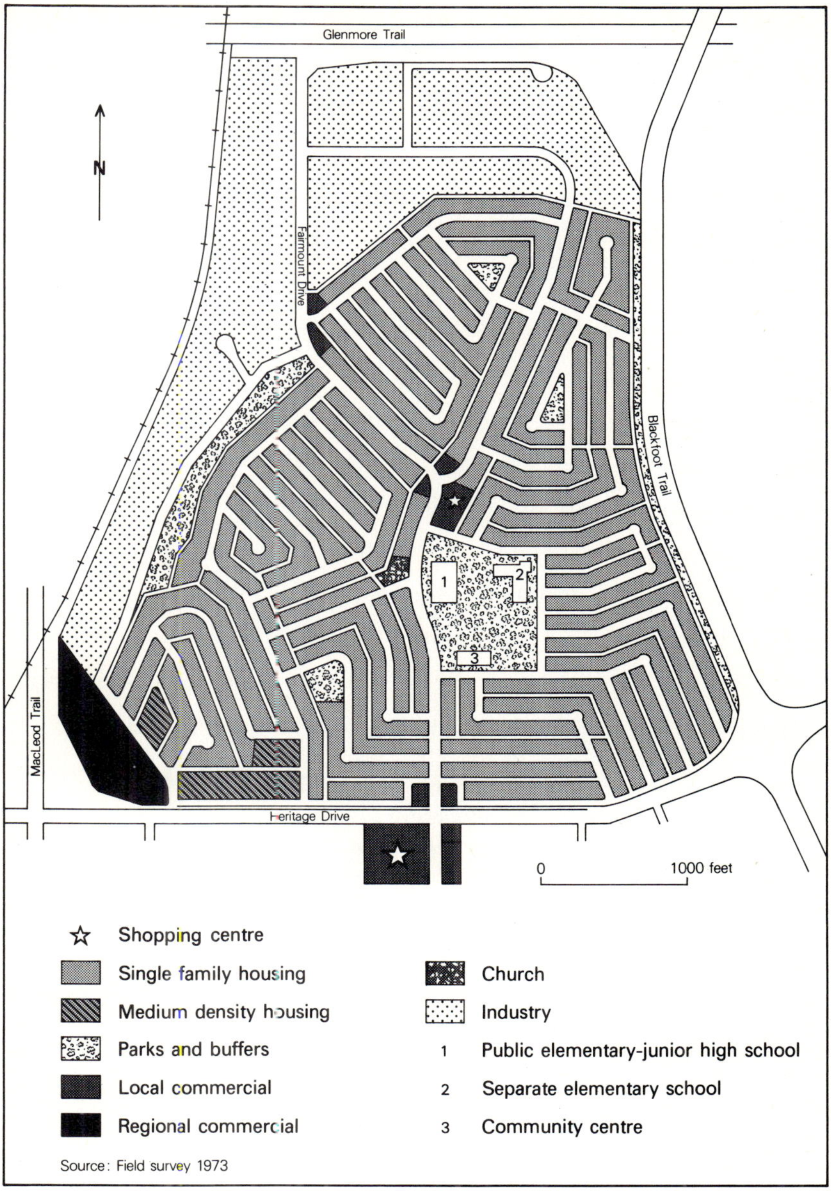

FIGURE 4,5 Fairview, a planned neighbourhood unit.

tion.[24] The planners had no clear guidelines, let alone standards, to work from, in determining the sizes of shopping centre sites. The developers were equally inexperienced, but the technical planning board was always wanting to believe that judgments of commercial land needs were best left to the market.[25] The public responsibility, in this view, was limited to technical matters such as street accesses and off-street parking.

This laissez-faire attitude towards the demand for commercial space created another problem that is illustrated in the Fairview neighbourhood unit. Just as there was no precise understanding of the relationship between individual shopping centres and their catchment areas, as the basis for determining their size and range of services, so there was no precise understanding of the impact of competitive centres on the spatial pattern of retail facilities. Competition was expected, of course, but the developers always presumed that it would operate in their favour. The Fairview planners offered exactly this argument in justification of their proposed shopping centre site:[26]

> Since this particular commercial district is on the secondary thoroughfare, it may also serve adjacent subdivisions to the south, therefore making its operation more efficient and large enough to be of better service to the district in which it is situated.

The principle may have been sound but it was written about the wrong shopping centre. A more convenient site, where the arterial street separating the Fairview and Acadia neighbourhood units (Heritage Drive) intersects with the major internal collector street (Fairmount Drive), was preferred by a supermarket chain and drained away the ancillary services that might otherwise have been located in the Fairview centre.

In addition to uncertainty about the size and spacing of neighbourhood shopping centres, it is evident from the above quotation that the planners were also uncertain about their locations. This, in turn, reflects Calgary's one striking departure from Perry's neighbourhood unit model. In place of an internal street system that excluded any form of through traffic (as a safety and privacy measure) Calgary adopted a system of secondary or collector streets that extend continuously through series of neighbourhood units. In Perry's model, all through traffic is diverted around the neighbourhood units; in Calgary, much of it is carried by the internal collector streets, such as Northmount Drive or Fairmount Drive. One of the consequences has been pressure for commercial land use along the collectors, almost, on occasions, to the extent of ribbon development. In Perry, the commercial sites are restricted to the perimeter of the neighbourhood unit; in Calgary they are not. Although the major local shopping centres have been placed at the points-of-entry to the neighbourhoods, reflecting the entrepreneurs' perceptions of maximum accessibility, the planners were ambivalent. Fairview is a case in point. On the one hand, Fairmount Drive was deliberately given a sinuous course to discourage through traffic;[27] on the other, its status as a secondary route, serving more than the Fairview neighbourhood, was offered as justification for the interior location of the Fairview shopping centre. Moreover, the planners could not resist the temptation to distribute small commercial sites along Fairmount Drive. In theory, these were to complement the neighbourhood centre by providing lower-order services for walk-in customers; in practice, they further undermined the attractiveness of the principal site. Once again, the planners were displaying their own uncertainty about the demand for retail space at the neighbourhood scale.

In general, the technical planning board's response to the locational problem was the same as for the size and spacing problems: let the market decide. In one of its rare policy statements, in 1955, the board ruled that commercial sites were to serve passing traffic as well as neighbourhood catchments, and both demands were to be allowed for in making decisions on location and size.[28] This meant that any site fronting on a through street had the potential to become a commercial site. Although the board occasionally voiced its opposition to continuous ribbon development in new residential areas,[29] it nonthe-less was prepared to permit a very close distribution of commercial sites. Major intersections were particularly vulnerable, as can be seen in Fairview (Figure 4,5). The board commonly granted commercial zoning to all four corners of an intersection, apparently anticipating a mix of local convenience shops and drive-in services. In fact, the greatest demand was for service stations, which were also the most controversial form of retailing in residential areas. From 1955 there were frequent complaints from home-owners who found that they were about to acquire a service station as a near neighbour.[30] Some professed complete ignorance of the commercial zoning; others believed that small stores would be permitted, but not service stations. In both cases, the board took the view that the problem was one of communication rather than location. Instead of ruling that conflicting land uses should not be located in proximity to each other, the board ordered stronger measures for advertising commercial zoning in advance of development.[31] Not until about 1960 did it begin to show concern for the numbers of service stations along some collector streets, notably Northmount Drive, and then only because the service stations were pre-empting sites that had been intended for local convenience shops.[32] The question of numbers in relation to need was openly declared to be none of the board's business, and local protests on amenity grounds were powerless to stop development for which some entrepreneur perceived a demand.[33] The result was an over-supply of commercial sites along collector streets. Many remained vacant for years, and may still be only partially developed; others were developed early, only to be abandoned later or converted to other retail uses. The mortality of service stations has been particularly high.[34]

It is readily apparent, then, that the second of Calgary's planning periods was marked by a great deal of misjudgment about the provision of commercial space. Despite the seeming assurance of a widely-accepted concept of residential planning, questions of the size, spacing and location of retail facilities were as ill-answered as they had ever been. The neighbourhood unit concept, of course, did little more then establish the principle that convenience retailing should be accommodated in the neighbourhood plan. It provided no hard-and-fast rules for the detailed decisions that had to be made. Decisions were therefore taken on an ad hoc basis, in the spirit of interim development control, each decision and its effects adding to the accumulation of knowledge and insights that were supposedly moving the planners towards the general plan for Calgary. The point was alluded to frequently,[35] and could have been sufficient in itself to justify a laissez-faire approach: how better to find out what was needed in new residential areas than by letting the market find its own level? The argument was never presented in planning terms, though. It was a matter of ideology that the technical planning board should not interfere with "the competitive spirit".[36] As a consequence, planning for commercial land use in new residential areas was limited to setting aside ample space in the most favoured locations. Whether development occurred, or the plan was modified, was left to the market to determine. The result was over-supply, of both land and retail services.

On the credit side for the planners, the years from 1954 to 1963 saw great changes in marketing behaviour. Developers, retailers and consumers alike were adapting to the new affluence, the new mobility and the new technology of convenience marketing, and it is little wonder that there was no clear understanding of their eventual effects. By the early 1960s, however, the pattern was becoming more stable. The measurement of need was no more precise, but the trend to unified shopping centres, requiring fewer but larger sites, was fully accepted. Even service stations, said the technical planning board in 1962, should be incorporated into the new shopping centres.[37] It was also coming to be appreciated that individual neighbourhood units could not be designed in isolation, and that the neighbourhood was not necessarily a self-contained service unit for even low-order retailing. The stage was being set for the change of planning scale that was to characterize the next period in Calgary's development.

THE PERIOD OF THE SECTOR CONCEPT, POST - 1963

The new conception of residential planning was made explicit in Calgary's first general plan which was adopted by city council in 1963. Among the schedules to the plan were statements of development principles for four large areas known as sectors, which ranged in size from 942 to 3,185 acres.[38] In effect, as the municipal land supply came closer to exhaustion, and the emphasis in land development shifted from the public sector to the private, the city planning officials also shifted their emphasis, from the details of subdvision design to schematic generalities. In this new perception of planning, the public responsibility was for the global pattern of development, and for the principles and policies by which it could be implemented; the filling-in of the detailed pieces was a private responsibility, though subject to governmental scrutiny to ensure their accord with the overall conception. The logic was undeniable, and had been obvious to the city planners for many years. The first versions of the sector plans were being drafted in the late fifties, although they were not actually approved until 1961.[39]

The best conceptual exposition of the sector planning concept was by a senior official of C. M. H. C., Humphrey Carver, whose book, Cities in the Suburbs, was published in 1962. There is no direct evidence of Carver's influence in the Calgary plans, but statements like the following could certainly have been drawn from the Calgary experience:[40]

> The population that supports one elementary school is not large enough to support the principal social and commercial services on which suburban life depends. The populations of three or four neighbourhoods together provide a reasonable economic demand for these services: the shopping centres, places of gathering and entertainment, and clusters of apartment houses. It is just these features of a city's living areas that are generally considered alien to family neighbourhoods . . .
> A comprehensive system for residential areas must, therefore, be conceived at a scale larger than the neighbourhood, so that it can embody all the essential features of suburban living which are extra-territorial to the neighbourhood.

Sector planning was thus conceived to have two basic motives: to provide a general framework into which specific development proposals could be fitted; and to create a community service unit larger and more realistic than the Perry neighbourhood. The neighbourhood did not disappear in this concept, but it became a sub-unit of a larger entity. Some services, such as elementary education, would continue to be organized at the neighbourhood scale; others would appear only at the community scale. To give identity and shape to the community, it would be organized around a major service node called a town centre by Carver and a sector centre by the Calgary planners. This node would include a shopping centre, high schools, libraries, churches, social service clinics, fire and police stations, and entertainment facilities.

Despite this background, however, the sector concept did not come into immediate full flower in Calgary. Several years of further refinement were needed, primarily to clarify the routines which the development industry had to follow. The eventual result was a four-step sequence.[41]

1. The design brief - a highly schematic plan for a sector or a large part of a sector. It is prepared by the city planning department and sets our the generalized land use and circulation systems, with particular emphasis on those features of suburban development that are public responsibilities (e.g. schools, parks and through streets).

2. The conceptual plan - also schematic but for a smaller area. It is the developer's opportunity to generalize about his land use expectations, within the constraints of the design brief.

3. The outline plan - in which the developer blocks out the land use zones and circulation pattern in greater detail.

4. The subdivision plan - the final detailed design.

The particular relevance of this sequence is that commercial land use planning is seen as a private responsibility. Apart from the sector centre, retail space needs are not identified until the stage of the conceptual plan.

Shopping Facilities: the Glenmore Case

The area south of Glenmore Lake was one of the first in which sector planning was attempted. The design brief was not accepted by city council until 1973 but the earliest version was completed in 1961,[42] and the first major subdivision, Braeside, was approved in 1962.[43] The sector area is 2050 acres and the target population is 42,000, based on an estimated gross density of 20.4 persons per acre.[44] Population projections, of course, are notoriously unreliable, and Glenmore is no exception. The original target, in 1963, was 30,000; the highest, in 1970, was 48,800.[45] Changing conceptions of the likely density of development and population structure were largely to blame, which means, in turn, that projections of community service needs are thrown into doubt. In particular, if the ultimate population figure is so uncertain, how can something as difficult as retail space demands ever be forecast in advance of development? It is not surprising that the city planning authorities should have disclaimed the responsibility.

The strongest statement about shopping facilities in the Glenmore sector appears in the 1963 general plan, where there is a suggestion that a population of 30,000 should support two shopping centres on sites of 10 to 12 acres. This appears to be completely intuitive. There is no attempt in the plan to establish a rational basis for retail forecasting, or even to discuss the principles of commercial land use planning within sectors.[47] The vagueness is underlined by the intended distinction between the two Glenmore centres. One was described as the "neighbourhood" centre and community focus, meaning the sector centre. It was given a central situation, at the major intersection within the sector, and this has persisted into the design brief (Figure 5,5). The other was located at a less important intersection and was expected to accommodate a "local" (that is neighbourhood) centre. In fact, this was scaled down in 1968 to a site of 4.5 acres to permit the construction of a standard neighbourhood shopping centre, anchored on a supermarket.[48] It was also recognised that other local facilities would be needed as development proceeded, but no attempt was made to pin them down.

In 1968, the municipal planning commission produced a revealing policy statement when reviewing the subdivision plan for the Oakridge neighbourhood in Glenmore.[49] The following criteria, it was decided, had to be considered in evaluating shopping centre proposals: the type of thoroughfare and intersection; the number of people to be served;

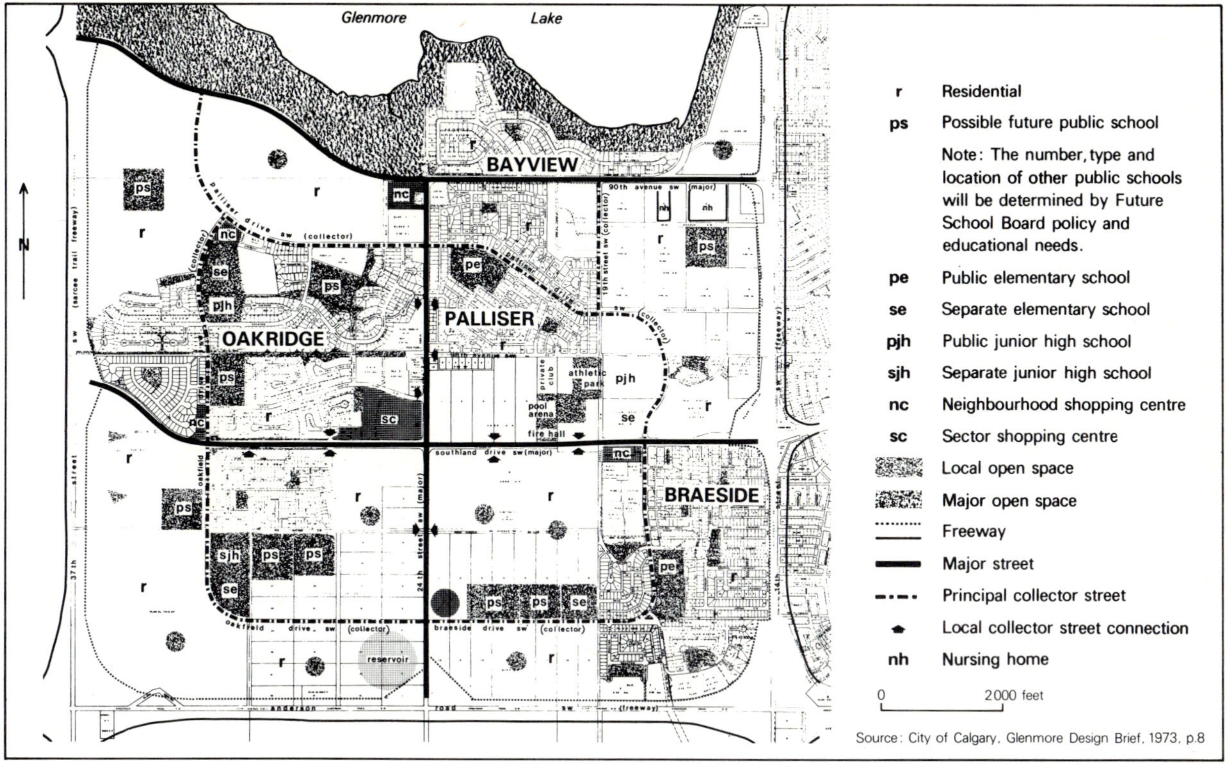

FIGURE 5,5 Glenmore Design Brief: general land use and circulation systems.

the intended size and function of the centre; and the need to allow equality of opportunity for all developers in the sector to build shopping facilities. In effect, the commission was reiterating the laissez-faire philosophy of its predecessor, the technical planning board. It was up to the entrepreneurs to decide how many shopping centres there should be, of what size and function. The commission's perceived responsibility was to rule on site selection with the aim of maximizing customer convenience (and hence the drawing power of the centre) while minimizing the disruption of traffic. In the Oakridge case, for example, the commission approved the neighbourhood shopping centre but changed its location.

An even more striking demonstration of the rule of the market place, from almost the same time, concerned the development of two regional shopping centres only a mile apart in northwest Calgary. The issue became both confusing and protracted, but the basic argument was simple: if a regional shopping centre specializes in high-order retail services, and requires a trade area of 150,000 to 400,000 people to survive,[50] does it make any sense to allow two of them to be built close together? In the event, both projects were approved and have come under development. The planning commission was explicit in its justification:[51]

> It is not uncommon for a regional centre to develop with two department stores and in a relatively short time, another centre materializes within a mile . . . This then accommodates major stores which were unable to locate in the other centre. This is the essence of the free enterprise competitive system.

In an attempt to clarify the principles of commercial land use planning, in preparation for the revision of the general plan, the city planning department produced two reports in 1968 and 1969. The first was primarily a classification and inventory of shopping centre development in Calgary, with some reference to more general studies by the Urban Land Institute.[52] Seventy-three shopping centres were identified, all of them built in a span of fifteen years. The report's main contribution was to group each of these centres into a four-tiered hierarchy, which was to form the basis of the classification scheme adopted in the 1970 and 1973 general plans (Table 1,5). The principal difference is that the report also emphasised the popularity of another type of local convenience centre, smaller than the neighbourhood centre and lacking a supermarket (the Fairview type). Particular stress was placed on the fact that over half of the 33 centres in the group had been built since 1961. Does their omission from the general plans mean that they had fallen out of favour with the planners?

The second report is a working paper with the title, A Method for Estimating Requirements for Neighbourhood Shopping Centres in New Areas.[53] It is based on the observation that an over-supply of neighbourhood retailing is "poor planning," but that the developers of shopping centres cannot be depended upon to carry out the feasibility studies that are ideally required. Estimations of need, it is therefore implied, must be a public responsibility, and the body of the report is a description of a possible technique, based on population and income parameters. Calgary's planning authorities, however, have yet to be convinced. The revised versions of the general plan, in 1970 and 1973, make no reference to either principles or standards for determining effective demand.[54] The only standards for shopping centre development refer to site area and parking; the only principles concern location in relation to thoroughfare type and the physical relationship of the centre to its environs.

This interpretation of the current perception of public responsibility in planning for sector retailing is reinforced by the minutes of the municipal planning commission and by the design briefs. In Glenmore, for

TABLE 1,5

PLANNED SHOPPING CENTRE TYPES
AND CHARACTERISTICS

TYPE	CHARACTERISTICS	LOCATIONAL CRITERIA
Regional Shopping Centre Provides shopping goods, general merchandise, apparel, furniture and home furnishings in depth and variety. Comes closest to reproducing shopping facilities once available only in central business district.	Principal tenant-major department store(s). Acreage - minimum of 30 acres. Gross leasable area - exceeds 300,000 square feet.	At the junction of an expressway and major street or two major streets and desirably along mass transit corridor.
Sector Shopping Centre Provides, in addition to "convenience goods," some selection of other goods such as apparel and hardware as well as banking, professional and recreational services.	Principal tenant - junior department store(s). Acreage - minimum of 8 acres. Gross leasable area - 90,000 square feet.	At the junction of two major streets or major street and expressway.
Neighbourhood Shopping Centre Provides for the sale of "convenience goods," i.e. daily living goods such as food, personal service and drugs.	Principal tenant - usually a supermarket. Acreage-average of 2.48 acres, range 0.64 to 6.13 acres. Gross leasable area-average 27,600 square feet.	Preferably at junction of a major street and a collector or two major streets.

Adapted from The Calgary Plan (1973), p. 7.5.

example, the sector plan (Figure 5,5) allocates four neighbourhood centres, but these are the only ones for which final subdivision plans have been approved, on the initiative of developers. Others will undoubtedly be added, as later neighbourhood designs are advanced, but the initiative will continue to rest with the developers. Apart from the site principles that were agreed upon in the Oakridge case, the municipal planning commission has been motivated by only one clear concern, which is to protect the prospective viability of the sector centre as the locus of more specialized retail services. Thus, no neighbourhood centre can be approved closer than one half mile from the intended site of the sector centre. If possible, it should be on a thoroughfare with through traffic that is not passing the sector centre: the site of the Oakridge neighbourhood centre, for instance, was shifted north from Palliser Drive to 90th Avenue in the hope that it would tap through traffic and so be less dependent on the sector population.[55]

The emphasis on the sector centre, of course, is fundamental to the sector planning concept. The notion of the sector as a community service unit depends on there being a focal node at which the higher-order services are concentrated. If the retail segment of these services is undermined, the rest of the notion is also likely to crumble, and the fine sense of hierarchical order, on which the whole concept is built, will become so much dust. In fact, in the Glenmore case, the sector centre may already have been critically weakened by decisions in the public service sector. In the 1963 plan, a combined senior high school and swimming pool were included on a site adjacent to the sector shopping centre. By 1973, the school board had decided that a high school would not be needed. Moreover, the design brief designates another large site, half-a-mile east of the sector centre, for a complex of swimming pool, arena, athletic park, branch library, health clinic, social service centre and firehall. Because of a problem of land acquisition, Glenmore seems likely to end up with a binodal sector centre, one for public services and the other for retailing.

From this example, it is evident that the change of planning scale does not make it any easier to implement the community service aspect of residential development. In some respects, it is actually more difficult than under the neighbourhood unit concept. There are problems of assembling large central sites for public services, and the whole question of phasing becomes much more difficult to resolve. Because of the cost of developing a sector centre, for example, it is not likely to be built until the housing areas are almost fully occupied. But this might mean that neighbourhood centres are over-developed to provide a reasonable level of service in the years before the sector centre is built. Will they continue to be viable afterwards? The Calgary planners are unable to answer this question, which illustrates, yet again, their persistent uncertainty about the demand for retail services. In the final analysis, the market still decides.

CONCLUSION

The central conclusion of this case study is inescapable: even under a planning system that demands a great deal of public involvement in the land development process, it is difficult, and perhaps impossible, to supplant the role of the market in determining the spatial pattern of retail facilities in suburban areas. Forecasting is a parlous task at the best of times, and the demand for retail services is probably subject to as great a range of variables as any element of the urban system. Furthermore, even if demand can be established, the viability of a particular form of retailing is by no means assured; a whole new set of factors, such as management policies and customer response, comes into play. It would be unreasonable to expect urban planners to be able to forecast the consequences with any great nicety, and in Calgary they have not even tried. Judgments of demand have been arrived at intuitively, and then only in the sense of ensuring that adequate space is set aside to allow the entrepreneurs to respond to the perceived demand without serious encumbrance. The results, as the Calgary example has shown, are far from perfect. Adjustments in response to competition and conflict have been made repeatedly, and are as likely to occur under sector planning as under neighbourhood unit planning.

On intellectual grounds, this may seem to be an irrational procedure, antithetical to the spirit of planning. On pragmatic grounds, though, it is entirely realistic. In a society in which competitive enterprise is a fundamental value, it would be futile for urban planners to attempt a high order of manipulation. In Calgary, they have restricted themselves to those aspects of commercial development that can be accepted as public responsibilities of the physical planner. This means such things as site size, parking allocation, relationship to the thoroughfare system and impact on nearby development.

On the positive side, the dedication of Calgary's planners to the concept of the community service unit has been consistent for twenty-five years. If nothing else, it has given developers and retailers the assurance that their needs were being taken seriously. Neighbourhoods and sectors may never be effective economic or social units but they are certainly physical ones, and local retailing is completely accepted as one of their essential elements. A division of responsibility has therefore emerged: the entrepreneurs determine how many shopping centres there will be, of what size and function; the city planners determine how they will fit into the residential areas that they are expected to serve.

In terms of the impact that this evolving process has had on the spatial pattern of retail facilities in Calgary, an increasing degree of geometric order is apparent. In the period of modified grid development, commercial sites were small and closely spaced; they were highly localized supplements to the central business district. Under the neighbourhood unit concept, which coincided with the first decentralization of retailing, the small convenience sites came to be mixed with the larger sites of neighbourhood shopping centres. They followed a strongly linear arrangement along the major collector streets that linked the neighbourhood units in series. Now, under the sector concept, a pattern closer to the classic central place pattern is being moulded. The sector centre has its cluster of satellite neighbourhood centres grouped regularly around it, and together they form the lowest levels of a four-tiered hierarchy.

REFERENCES

1. See, for example, KEEBLE, L. Principles and Practice of Town and Country Planning. London: The Estates Gazette, second edition (1959), pp. 162-173; American Public Health Association, Committee on the Hygiene of Housing. Planning the Neighbourhood. Chicago: Public Administration Service (1960); and Central Mortgage and Housing Corporation, Site Planning Handbook. Ottawa: C. M. H. C. (1966).

2. The early development of Calgary has been discussed in a number of previous papers: SMITH, P. J. "Calgary: A Study in Urban Pattern," Economic Geography, Vol. 38 (1962), pp. 315-329; SMITH, P. J. "Change in a Youthful City," Geography, Vol. 56 (1971), pp. 1-14; and SMITH, P. J. "Changing Forms and Patterns in the Cities," in SMITH, P. J. (ed.) The Prairie Provinces. Toronto: University of Toronto Press (1972), pp. 99-117.

3. GERTLER, L. O. "Some Economic and Social Influences on Planning in Alberta," Plan Canada, Vol. 1 (1960), pp. 115-121.

4. "Planning Association Honors Calgary Man," The Albertan, November 1, 1950.

5. See, for example, an editorial in The Albertan, August 29, 1949.

6. Letter from J. Ivor Strong, City Engineer, to J. M. Miller, City Clerk, November 9, 1949.

7. Bylaw 4237, A Bylaw of the City of Calgary to Appoint a Technical Planning Board, August 20, 1951.

8. Province of Alberta, Interim Development Order, City of Calgary, October 10, 1951.

9. Bylaw 4271, A Bylaw of the City of Calgary to provide for the interim control of development during the preparation of the General Plan for the development of the City, March 31, 1952.

10. Letter from V. A. Newhall, City Commissioner, to G. M. Brown, chairman of the Town Planning Commission, October 4, 1945.

11. The Albertan, March 5, 1947.

12. See, for example, The Albertan, November 4, 1949, and January 13 and February 15, 1950; Acreage for Parks and Playgrounds, a report from the Town Planning Engineer to the Town Planning Commission (1950); and Calgary Town Planning Commission, Minutes, April 20, 1950.

13. The Albertan, March 10, 1948.

14. MARTIN, A. G. Progress Report on the Comprehensive Plan. Calgary Town Planning Commission, August 23, 1951.

15. PERRY, C. A. "The Neighbourhood Unit," in Neighbourhood and Community Planning: Regional Survey of New York and its Environs, Vol. 7. New York: Committee on Regional Plan of New York and its Environs (1929).

16. MARTIN, op. cit., p. 5.

17. Letter from J. M. Miller, City Clerk, to A. G. Martin, City Planner, June 9, 1953.

18. Letter from J. A. Lamb, Town Planning Engineer, to J. M. Miller, City Clerk, March 16, 1950.

19. Calgary Technical Planning Board, Minutes, January 22, 1958. June 8, 1955; and MARTIN, A. G. "Land Development in Calgary," Habitat, Vol. 5 (1962), p. 19.

20. See, for example, Calgary Technical Planning Board, Minutes, January 22, 1958.

21. Government of the Province of Alberta, The Subdivision Regulations, O. C. 969-53.

22. See, for example, Calgary Technical Planning Board, Minutes, March 5, 1961.

23. DAVIS, E. H. and GORDON, G. E. "Fairview - A Planned Development," Community Planning Review, Vol. 8 (1958), pp. 130-136.

24. See, for example, Calgary Technical Planning Board, *Minutes,* September 19, 1962.

25. See, for example, letter from A. G. Martin, City Planner, to C. B. Cummer, City Clerk, June 4, 1959.

26. DAVIS and GORDON, *op. cit.,* p. 133.

27. *Loc. cit.*

28. Calgary Technical Planning Board, *Minutes,* September 22, 1955.

29. See, for example, Calgary Technical Planning Board, *Minutes,* March 15, 1961.

30. Calgary Technical Planning Board, *Minutes*, June 15, 1955 and June 5, 1963; and letter from C. B. Cummer, City Clerk, to A. G. Martin, City Planner, August 2, 1955.

31. Calgary Technical Planning Board, *Minutes*, September 27, 1955, and November 19, 1958.

32. Calgary Technical Planning Board, *Minutes,* June 29, 1960.

33. See, for example, Calgary Technical Planning Board, *Minutes,* June 15, 1955.

34. Northmount Drive was always the principal subject of concern. In a field survey in 1973 two service stations and a drive-in restaurant were found to be abandoned.

35. See, for example, Calgary Technical Planning Board, *Minutes,* January 5, 1955.

36. Calgary Technical Planning Board, *Minutes,* June 29, 1960.

37. Calgary Technical Planning Board, *Minutes,* February 14, 1962.

38. *Bylaw 5997,* A Bylaw of the City of Calgary to adopt a General Plan for the development and use of land within the City of Calgary, August 19, 1963 (Group G. "Sector Plans").

39. Calgary Planning Advisory Commission, *Minutes,* September 28, 1961.

40. CARVER, H. Cities in the Suburbs. Toronto: University of Toronto Press (1962), p. 60.

41. Calgary, City Planning Department, The Design and Processing of Subdivisions in Calgary. City of Calgary (1973).

42. Calgary Planning Advisory Commission, Minutes, April 20, 1961.

43. Calgary Technical Planning Board, Minutes, February 14, 1962.

44. Calgary, City Planning Department, Glenmore Design Brief, approved by City Council, May 28, 1973.

45. Calgary, City Planning Department, Design Brief of the Glenmore Sector (Preliminary Report), July 10, 1970.

46. Calgary, City Planning Department, Residential Development in Calgary: Inventory and Prospect. City of Calgary (1973).

47. Bylaw 5997, op. cit., Group D-2, "Commercial Development."

48. Calgary Municipal Planning Commission, Minutes, May 1, 1968.

49. Calgary Municipal Planning Commission, Minutes, July 24, 1968.

50. Calgary, City Planning Department, Shopping Centres in Calgary. City of Calgary (1968), p. 45.

51. Calgary Municipal Planning Commission, Minutes, December 6, 1967.

52. Calgary, City Planning Department Shopping Centres in Calgary. City of Calgary (1968).

53. Calgary, City Planning Department A Method for Estimating Requirements for Neighbourhood Shopping Centres in New Areas. City of Calgary (1969).

54. The Calgary Plan. A General Plan prepared on behalf of the civic administration by the City of Calgary Planning Department, March 1970, section 5, "Commercial Development;" and idem, May 1973, section 7, "Commercial Development."

55. Calgary Municipal Planning Commission, Minutes, July 24, 1968.

PLATE 22
The University of Calgary, a major determinant of growth in the northwest.

PLATE 23
The University of Calgary.

University of Calgary Pho

CHAPTER 6

FOOD STORE-DWELLING LINKAGES IN SELECTED AREAS OF CALGARY

Denis B. Johnson

University of Alberta

INTRODUCTION

This study deals with the interaction between inner-city residents, rich and poor, and food stores in Calgary. The concept of the linkage is used as a descriptive tool to generalize about this interaction. The linkage is defined as:

> The relationship between establishments characterized by frequently recurring interaction. Linkage leads to the movement of persons and goods between linked establishments and generates a tendency on the part of establishments to seek proximate locations or locations that are mutually accessible.[1]

Food store-dwelling linkages have three components:

1. the dwelling,
2. the connection, and
3. the food shop.

Description of the dwelling reveals information about the food shopper. The connection component describes how the interaction between dwelling and store takes place. Analysis of the spatial distribution of food stores provides the framework within which food shopping takes place. The purpose of this study is to analyse the spatial patterns of food stores in the southern sector of Calgary, and to describe the spatial relationships between dwellings and food stores in two inner areas of the city.

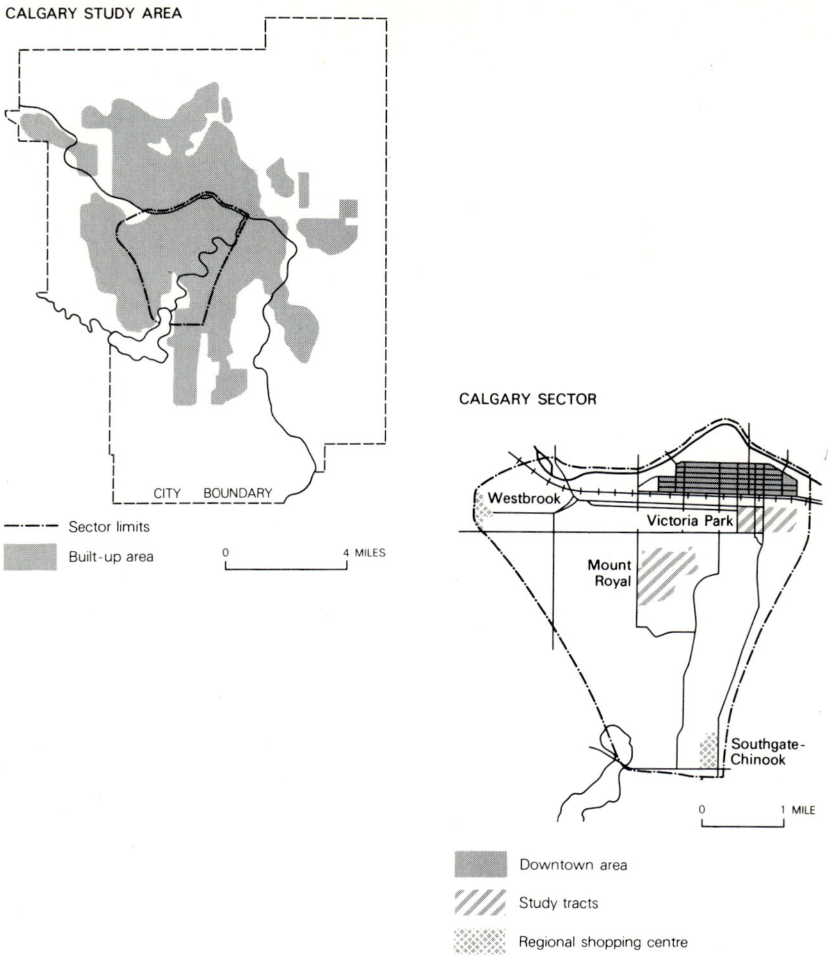

FIGURE 1,6 The study area.

Field surveys provided most of the data for this study. Field mapping provided information on the form, function, and spatial distribution of food stores in the study area. Residential interviews of sample dwellings in two inner tracts provided data on the dwelling-connection components of the linkages.

THE STUDY AREA

The study area is made up of two spatial units in Calgary, the study tracts and the study sector (Figure 1,6). The samples of dwellings for the interview survey were drawn from two tracts, Victoria Park and Mount Royal. These areas represent two general types of residential areas found in western cities:

1. Victoria Park, a working class area, is relatively old, and is adjacent to the downtown area. The fair to poor condition of dwellings in this tract is assumed to reflect a population of food shoppers with low mobility and who rely on local shopping facilities; and

2. Mount Royal, a relatively old upper class area, is situated about one mile south of the downtown area. This tract is assumed to contain highly mobile shoppers who depend mainly on the automobile for access to food stores.

The study sector was delimited to include food stores which were most likely to be used by residents in the study tracts. The sector originates in the downtown area and extends beyond the tracts into the suburbs. Two regional shopping centres mark the limit of the sector to the south and west (Figure 1,6). A field reconnaissance indicated that no great improvement in food shopping facilities could be found beyond these centres.

FOOD STORE PATTERN AND FORM

The food store is normally considered a convenience or low order retail establishment. Consumers are unwilling to travel as far for food items as for clothing, furniture, gifts, and other shopping, or higher order goods. The automobile, however, has increased intraurban mobility making formerly distant locations more accessible. The refrigerator, and tinned and frozen perishables have improved the food storage facilities of dwellings. These developments have resulted in food shopping trips of greater distances, and of lower frequencies.

The greatest increase in mobility has occured in the middle to upper income areas of the city. Here, food shopping facilities, mainly supermarkets and food stores in planned shopping centres, cater to the automobile. Although there are remnants of pre-automobile shopping facilities in most of the inner city, the low income areas have a greater share of, and dependence on these establishments. Corner grocery stores, self-service grocery stores, and small specialized stores (butchers, bakery shops, green-grocers and fish stores) are found in, and adjacent to older low income residential areas. In the Calgary sector, these establishments are found mainly in and around the downtown area (Figure 2,6).

The changes in food store patterns and the consumers' reaction to these changes are summarized by a simple model of foodstore-dwelling linkages (Figure 3,6). Linkages in this model are illustrated according to their position on a continuum extending from pre-supermarket, pedestrian oriented patterns to a system which may see all food shopping done via telecommunications systems:

> In the not too distant future a substantial share of these (convenience) items may be marketed through a relatively few large central distribution facilities in each major market area. Consumers will never set foot inside these centers, except possibly for guest tours. Instead, retail transactions will be made by electronic telecommunications and push-button devices installed in private homes and hooked on-line to data processing networks.[2]

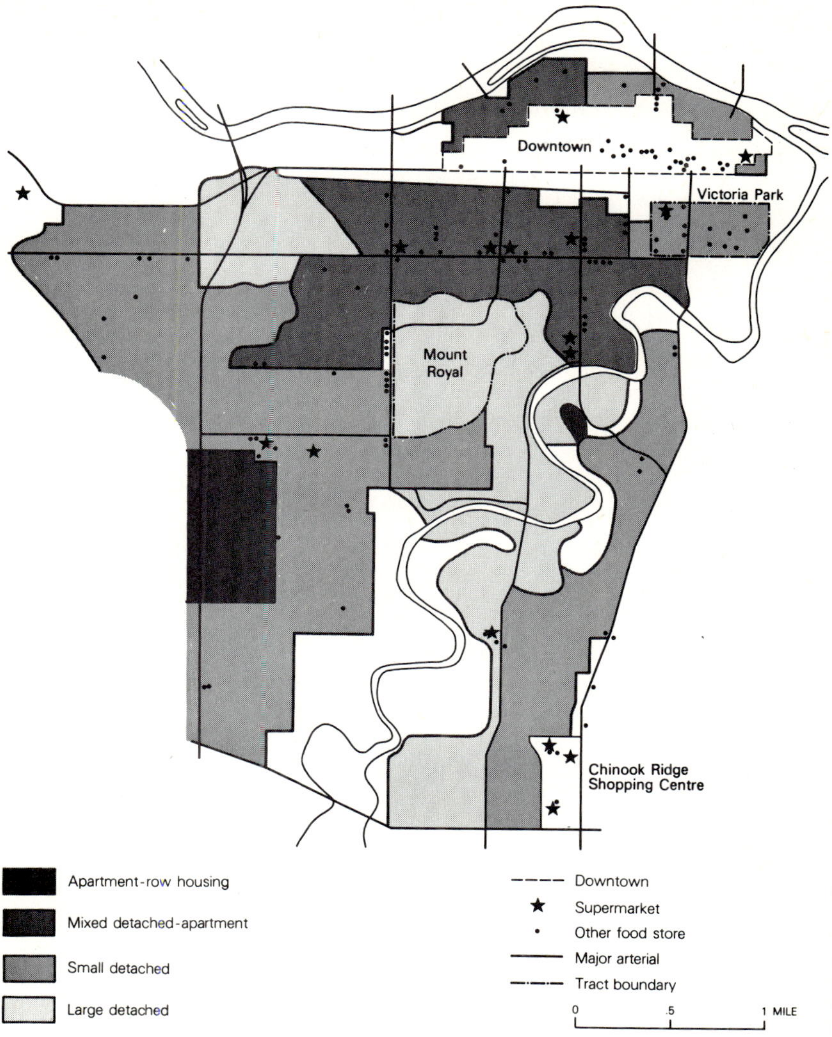

FIGURE 2,6 Food stores and housing types.

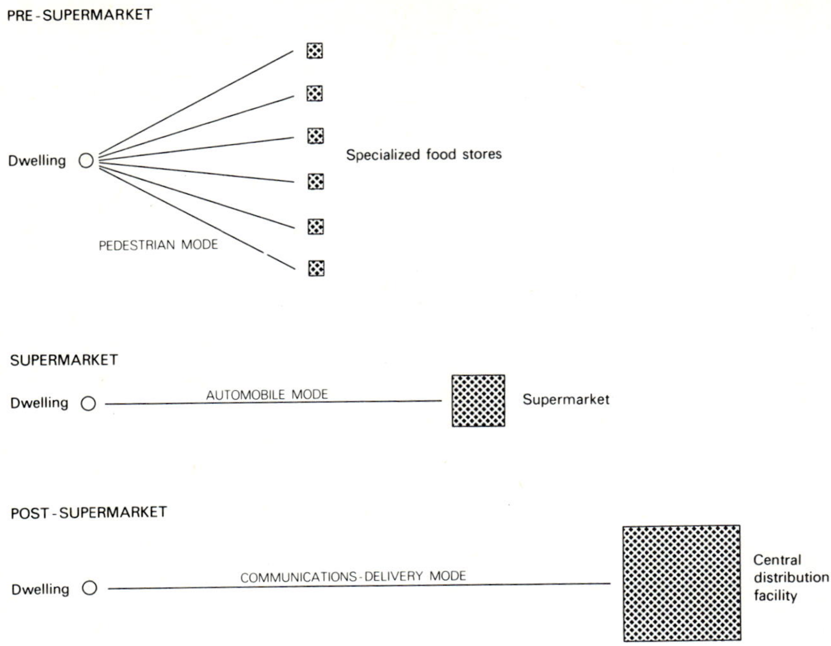

FIGURE 3,6 Continuum of change in food store-dwelling linkages.

Food Store Classification

The purpose of this classification is to provide a means of describing the functions of individual food shops, thereby providing a comparison of food shopping facilities available to residents in the study tracts. The classification system, the result of field observations in the study sector, is made up of the following types of food shops:

Supermarkets. A supermarket must have at least three checkouts, a sales floor area of more than 2,000 square feet, and provide shopping carts or baskets. The range of merchandise must include all food groups including fresh meat and fresh fruits, and vegetables. It must provide off-street parking or be located in a planned shopping centre which provides this facility.

Self-Service Grocery. A self-service must have at least one checkout and provide shopping carts or baskets. The range of merchandise must include all food groups including fresh meat and fresh fruit and vegetables.

Corner Grocery. The corner grocery is difficult to define. Any shop selling groceries which does not qualify for the preceding categories is classified as a corner grocery. In general, the corner shop appears to cater to:

1. the person who wants to pick up a few grocery items and does not want to spend time travelling to a major food centre (or the time waiting in a queue at an adjacent supermarket for one or two items); or

2. the person who requires food items when other food shops are closed.

Corner shops, therefore, may have one or more of the following characteristics:

1. they are generally located within or adjacent to a residential area for maximum convenience;

2. they may be open during evenings and Sundays when other food shops are closed, usually in compliance with city regulations. In Calgary, although there are no restrictions on Monday-Saturday hours, most shops are open Thursday and Friday evenings and most corner shops are open every evening and on Sundays;

3. corner shops usually sell a small amount of non-food items, especially tobacco products and

4. they appear to sell more frequently needed and perishable items such as bread and milk.

Butcher. This type of shop specializes in selling fresh meat. Final cutting of the meat usually takes place in the shop.

Fish Store. This shop specializes in selling fresh fish and sometimes fresh poultry.

Bakery. The bakery specializes in selling bread and pastries, mostly unpackaged.

Greengrocer. This shop specializes in selling fresh vegetables and fruit.

Specialty Foods. This type includes any food shop which does not fall into the above categories. Types of shops observed in this group are usually delicatessens selling imported foods and cooked meats. Also included are shops specializing in tea and coffee, jams and biscuits, or health foods.

Perhaps the following quotation is relevant when reflecting upon the problem of food shop classification:

> So in the matter of small retail trade, we need to revise the meaning of our vocabulary. A dairy is a grocery that sells fresh milk. A bakery is a grocery that sells fresh bread, and butchers and fishmongers are, increasingly, groceries that sell fresh meat and fish. A chemist is a Thing shop that dispenses medicine. An ironmonger is either a hardware shop, or a plastic ware shop with groceries and Things. A grocery is a grocery and Thing shop. And a Thing shop is a Thing shop that doesn't yet sell groceries.[3]

Food Centres

Efficient personal shopping involves a consumer going to the nearest centre where she can purchase the food items she requires. This process is facilitated either by a grouping of a variety of food stores which are within walking distance of each other, or by a supermarket which provides most food items under one roof (Figure 3,6). In this study these forms of food shops are considered as food shopping centres by virtue of their pedestrian scale. The delimitation of such centres is based on an estimate of the maximum distance pedestrians will walk from one shop to another after arriving at the centre by car or other modes of transportation. Estimates of this distance depends on such variables as the physical condition of the individuals using the centre, the weather, the topography of the site, and the attitude of the individuals towards walking, not to mention that some parts of the urban environment are more suitable for walking than others.[4]

A number of studies have selected criteria for defining shopping centres. J. D. Nystuen[5] used the distance of 200 feet as the maximum distance allowable between shops in this delimitation of shopping centres while F. W. Boal[6] used the figure of 266 feet. Berry[7], in one of his earlier studies used an intuitive approach in defining centres pointing out that the groups of businesses should be spatially separated from other business uses by alternate types of land use. Berry and his co-workers have employed land value data to determine centres and their boundaries. This was done by selecting a minimum land value to define a centre and drawing the boundary of the centre at breakpoints in the land value profile across the centre. Simmons,[9] because of unavailability of land value data in Toronto, selected alternative criteria in this definition of shopping centres. These included presence of high order functions, upper floor uses, continuity of commercial uses, absence of low order functions and physical barriers. Finally, Jones[10] delimited retail centres in Belfast, on the basis of field observations and intuitive notions. From this outline it appears that there is little consensus as to what is the best definition of a shopping centre. F. W. Boal and the author[11] have also suggested that a definition of a shopping centre should include consideration of the spatial distribution of calls made by shoppers on multiple shopping trips. A shopping centre may not necessarily be a closely spaced grouping of stores, but a grouping spatially associated in terms of sets of calls made to them by shoppers.

The concept of the shopping centre is used in this study under the condition of excluding all establishments except food stores from a shopping centre area. This condition assumes that the primary purpose of food shopping trips is to obtain items sold by individual food stores or groups of stores such that it is possible to walk among stores in a given group. The term food centre is used to signify this modified form of the shopping centre and is defined as follows:

1. a food centre consists of one or more food stores one of which must be either a specialized food store, a supermarket or self-service grocery; and

2. the maximum distance between stores in a food centre does not exceed 250 feet walking distance.

FOOD STORES IN THE CALGARY SECTOR

Food stores in the study sector have been attracted to the down-town areas, to major arterials, and to some residential areas. Supermarkets make up an irregular pattern and occur mainly on arterials, sometimes isolated from other food stores, and sometimes in a cluster of food stores. There are two cases when two supermarkets are next to each other, and three supermarkets are located in the Chinook-Ridge Shopping Centre (Figure 2,6).

The downtown area contains a variety of food stores, especially the specialized and specialty types (Figure 4,6 and Table 1,6).

TABLE 1,6

FOOD STORES IN DOWNTOWN CALGARY

Type of Store	Number of Establishments
Supermarket	2
Self Service Grocery	2
Corner Grocery	6
Butcher	3
Fish Store	2
Bakery	3
Greengrocer	2
Specialty	11
TOTAL	31

Source: Field Survey, 1968.

Three forms of food centres are found in Victoria Park (Figure 5,6):

1. a small cluster of a butcher, bakery, corner grocery and a specialty shop (Danish delicatessen);

2. two supermarkets -- one appears with two other stores; and

3. an isolated self-service grocery, well within the tract.

Corner grocery stores are evenly distributed throughout the tract. The nearby downtown area adds to the variety of establishments accessible to Victoria Park residents.

The western boundary of Mount Royal is fringed by a number of food stores making up four centres (Figure 6,6). The largest centre offers a self-service grocery, a bakery, and a corner grocery store. Five centres are located to the north of the tract. Two supermarkets make up one of these, and a supermarket dominates the other. The remaining centres are made up of one or more specialized and specialty stores.

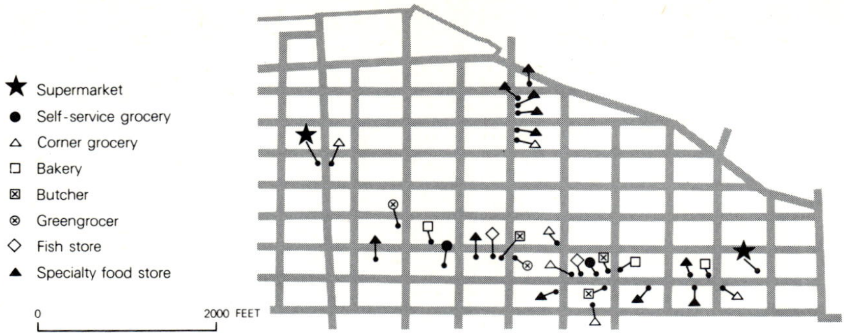

FIGURE 4,6 Food stores in downtown Calgary.

FOOD STORE - DWELLING LINKAGES

The Interview Survey

Although the locational patterns and forms of food shops are important components of the retail structure, the aspect of movement of people, goods, and information to and from these establishments is necessary for a fuller understanding of their function. Simmons points out that this pattern

> . . . is not haphazard but arises from a structure of retail activity, a system of regular interaction of customers and stores determined by economic and cultural values . . .[12]

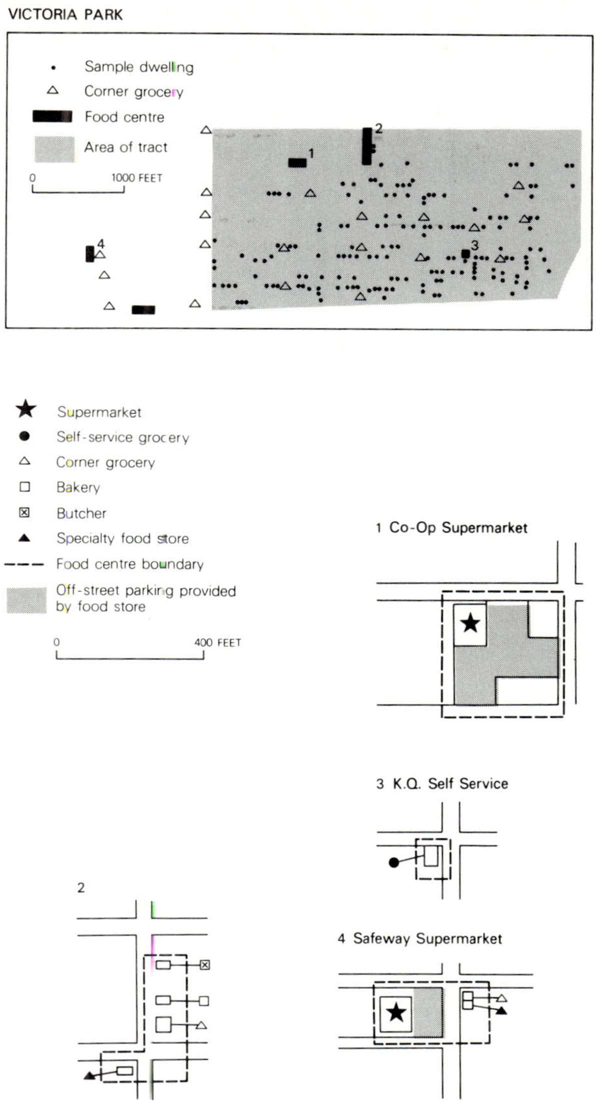

FIGURE 5,6 Main food stores in the vicinity of Victoria Park.

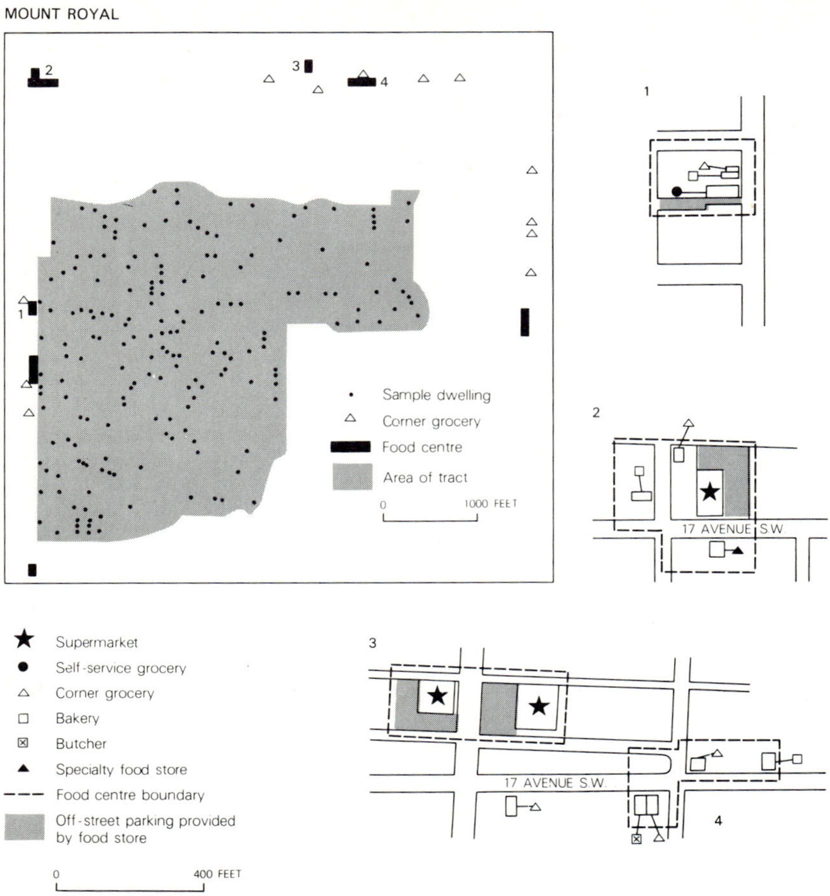

FIGURE 6,6 Main food stores in the vicinity of Mount Royal.

Residential interview surveys obtained information on tract resident characteristics relevant to food shopping, and on the nature of food shopping with linkage providing an operational framework. A simple random sample was selected from the population of dwellings in the study tracts. Sample units were dwellings as described in the 1966 Census of Canada.[13] The sampling frame consists of computer printouts of 1965 land use data and field checking by the author and interviewers. Sample size was based on available finances, and statistical tables.[14]

Interviewers made three attempts to obtain a response, in the morning, in the afternoon, and in the evening. A response was a successful interview with someone in the dwelling who could describe food shopping habits for the persons living there. If no response was obtained after three calls, the dwelling was dropped from the sample without replacement.

The interview schedule was designed to obtain information on the three components of the food shop - dwelling linkage, the dwelling, the store or stores frequented, and the connection between the dwelling and stores. Information on the food store used was obtained from responses to three sets of questions:

1. where they usually purchased seven specified food items -- meat, fish, pastries, bread, groceries, vegetables and fruit, and milk;

2. whether they used other food shops in addition to the ones listed under these food items; and

3. whether they made use of one or more corner grocery shops.

Information on the connection component was obtained by asking the respondent the main type of transportation-communication used in connection with all the shops listed, and the frequency of usage of each of these shops. Data on the dwelling component involved the access of the food shopper in the dwelling to a car, telephone, refrigerator, and deep freeze. In spite of the attempts to reduce the non-response, a fairly high rate of twenty-seven per cent was experienced in Victoria Park followed by eighteen per cent in Mount Royal (Table 2,6).

The analysis of the interview data was aimed at generalizing and comparing the spatial components of food shopping in the two tracts. Two approaches were taken: characteristics of the total accumulation of linkages for each tract are tabulated and compared; and data are tabulated to identify combinations of food stores associated with dwellings in each tract under the assumption that some dwellings respond to the

TABLE 2,6

SAMPLE POPULATION AND RESPONSE

	Victoria Park	Mount Royal
Number of dwellings	1602	668
Sample drawn	240	227
Response	174	186

retail pattern in a similar way (that is a number of respondents in a tract frequent the same store or groups of stores).

Accumulated Linkages

The Dwelling Component

The mobility of the food shopper influences her decision to frequent local food shops in the case of low mobility, or to travel further to more attractive shopping areas in the case of high mobility. Food storage facilities may influence the frequency of connections between dwelling and shop. Comparison of the four variables tabulated reveals Mount Royal as scoring highest in both mobility and storage indices.

TABLE 3,6

OCCURENCE OF CAR, TELEPHONE, REFRIGERATOR AND DEEP FREEZE IN STUDY TRACT SAMPLE DWELLINGS

	Victoria Park		Mount Royal	
	Number	Per Cent	Number	Per Cent
Car	91	52	178	96
Telephone	145	83	186	100
Refrigerator	166	95	186	100
Deep Freeze	45	26	96	52

Source: Interview Data, 1968

The Connection Component

The number of linkages per sample dwelling shows a considerable variation between the tracts (Table 4,6). Well over one-half of the Victoria Park dwellings had connections with two or less food stores. There was more intra-tract variation in Mount Royal, and the number of linkages per dwellings were generally greater.

Modes of Communication - Transportation

Car connections were the most prevalent in Mount Royal followed by home delivery linkages (Table 5,6). Although the automobile was used for about one-quarter of the Victoria Park food shopping, the pedestrian mode was the most frequently used. The home delivery mode in Mount Royal was used mainly for milk purchases. This mode was almost nonexistent in Victoria Park.

TABLE 4,6

NUMBER OF LINKAGES PER SAMPLE DWELLING
(Excluding home delivery linkages)

	Victoria Park		Mount Royal	
Number of Linkages per Dwelling	Number	Per cent	Number	Per cent
1	19	11	16	9
2	104	61	37	20
3	40	23	42	23
4	8	4	43	23
5	1	-	33	18
6	2	1	9	5
7	0	0	5	2
8	0	0	0	0
9	0	0	1	-
TOTALS	174	100	186	100

Source: Interview data, 1968.

TABLE 5,6

MODES OF TRANSPORTATION-COMMUNICATION

Mode	Victoria Park		Mount Royal	
	Number	Per cent	Number	Per cent
Car	102	26	529	66
Telephone	0	0	26	3
Home Delivery	4	1	141	18
Walk	257	64	54	7
Bus	17	4	9	1
Other	20	5	38	5
TOTALS	400	100	797	100

Source: Interview data, 1968.

Spatial Distribution of the Accumulated Linkages

The study sector was intuitively delimited to include food stores which were most likely to be used by residents in the two tracts. This delimitation is supported by the Victoria Park interview data as only one per cent of the linkages in this tract were outside the sector (Table 7,6). Even within the sector the main interaction took place in the areas adjacent to, or near the tracts (Figure 7,6). This adjacent area, termed the High Density Interaction Area, was arbitrarily defined as the area contiguous to each tract, where at least one-half of the linkages for that tract terminate. The boundaries of the High Density Interaction Area, hereafter abbreviated H. D. I. A., are shown on Figures 7,6 to 13,6. This area was selected to allow more detailed presentation of the linkages by adopting a larger scale of maps for the areas containing the highest density of linkages.

Four per cent of the Mount Royal linkages had store connections outside the sector, while about one half occured in the Mount Royal

H. D. I. A. Non-spatial linkages (involving telephone-delivery modes) were higher in the Mount Royal tract.

TABLE 6,6

SPATIAL AND NON-SPATIAL DISTRIBUTION OF LINKAGES

	Victoria Park		Mount Royal	
	Number	Per cent	Number	Per cent
High Density Interaction Area	372	94	395	50
Sector (excluding H. D. I. A.)	17	4	222	27
Outside Sector	7	1	31	4
Non-Spatial				
Home Delivery Agent	4	1	142	18
Other	0	0	7	1
	400	100	797	100

Source: Interview data, 1968.

The greater spatial complexity of Mount Royal linkages is shown by Figures 7,6 to 11,6. Figure 7,6 summarizes the linkages within the sector, but outside the H. D. I. A.'s. Mount Royal linkages dominated this map. Individual stores which are prominent on this map are Woodwards supermarket in the Chinook-Ridge regional shopping centre, a Safeway supermarket just west of Mount Royal, the Hudson's Bay grocery department and Billingsgate Fish Store in the downtown area, and even the Co-Op supermarket within the Victoria Park tract.

Car and Pedestrian Linkages in the H. D. I. A.'s

Differences and similarities in linkages are revealed by the desire line maps showing selected aspects of car and pedestrian linkages in the two H. D. I. A.'s (Figures 8,6 to 11,6). The car linkage maps (Figures 8,6 and 9,6) show the greater number and complexity of the Mount Royal

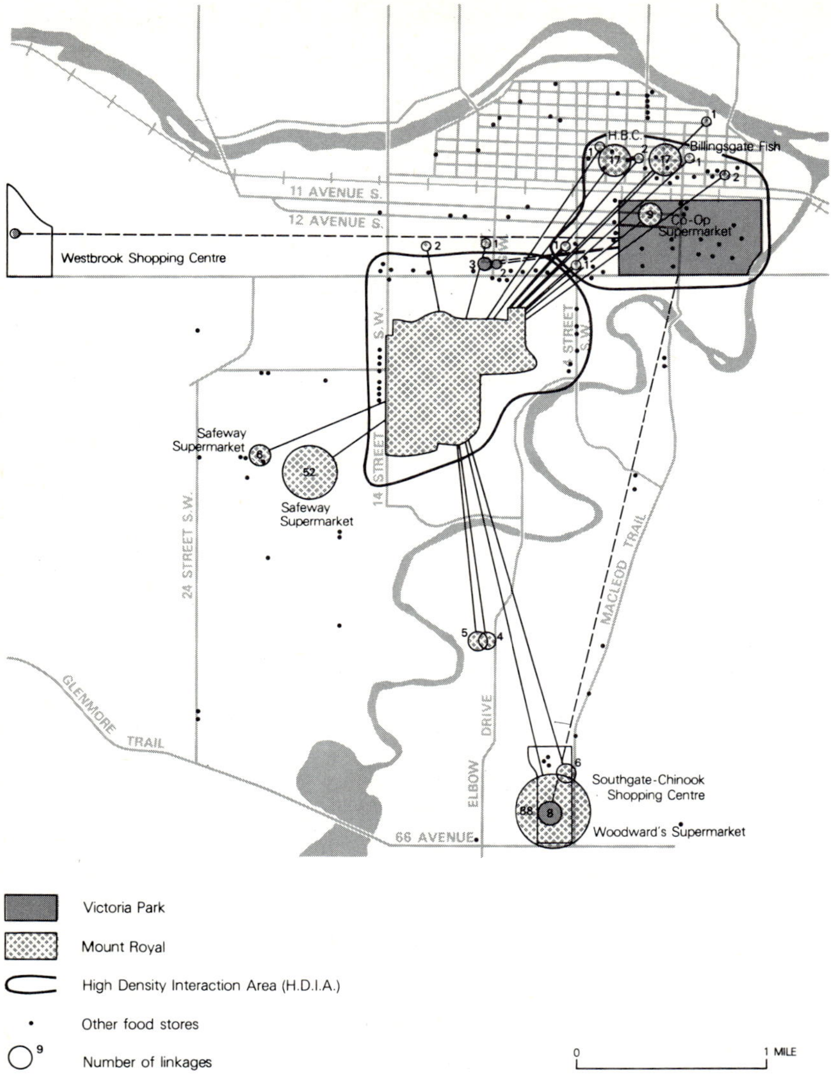

FIGURE 7,6 Linkages terminating outside the Victoria Park and Mount Royal Tracts.

linkages at this scale. In Victoria Park all of the car linkages were with either supermarkets or self-service stores. Car linkages to corner grocery stores, even for relatively short distances, were prevalent in the Mount Royal case but they were non-existent for Victoria Park (Figure 9,6).

Pedestrian linkages occured mainly in the western part of Mount Royal, and throughout Victoria Park (Figure 10,6 and 11,6). The lengths of pedestrian connections were similar in both tracts. This is especially evident when pedestrian connections with corner grocery stores occured (Figure 11,6). The range of this type of store appears to be similar, and the lack of pedestrian linkages throughout Mount Royal may reflect the fact that corner grocery shops occur only on the western boundary of the tract. Some respondents in Mount Royal were willing to walk as far, or further than those in Victoria Park, other respondents in Mount Royal took their cars even for short distances (Figures 9,6 and 11,6).

Combinations of Food Stores Frequented by Sample Dwellings 15

To identify combinations of stores frequented by one or more respondents, the interview data were tabulated to produce out of all the possible combinations of stores in the sector, three orders of combinations: individual stores, pairs of stores and triads of stores. Larger groupings were not tabulated because very few respondents frequent similar combinations of three or more stores. These tabulations were designed to answer the following questions: How do residents in each tract interact with food stores within the sectors? Are there individual, or larger groups of stores frequented by more than one sample respondent? This analysis was undertaken in order to test the assumption that food shoppers shop where all of their needs can be satisfied with the least effort. Supermarkets facilitate this requirement by providing a wide range of food items under one

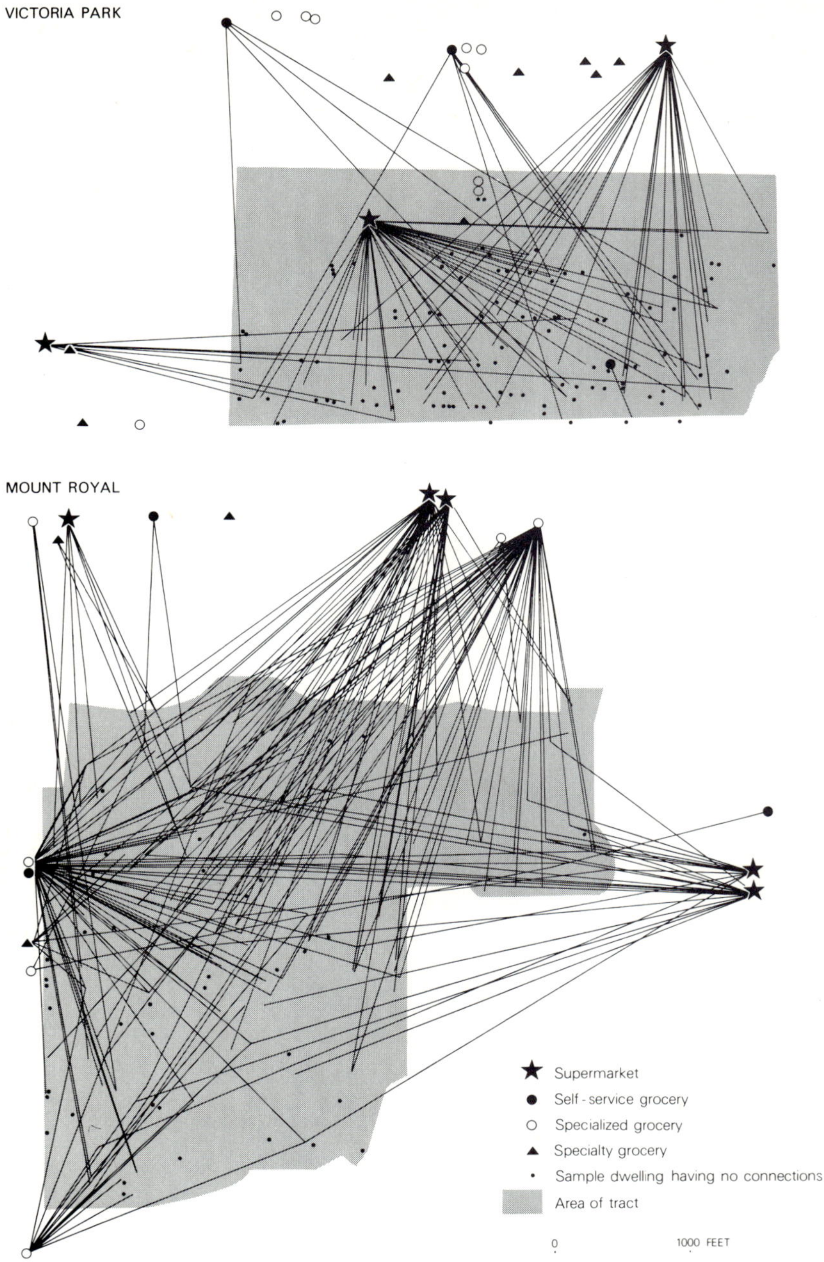

FIGURE 8,6 Car linkages - H.D.I.A. (excluding corner grocery shops).

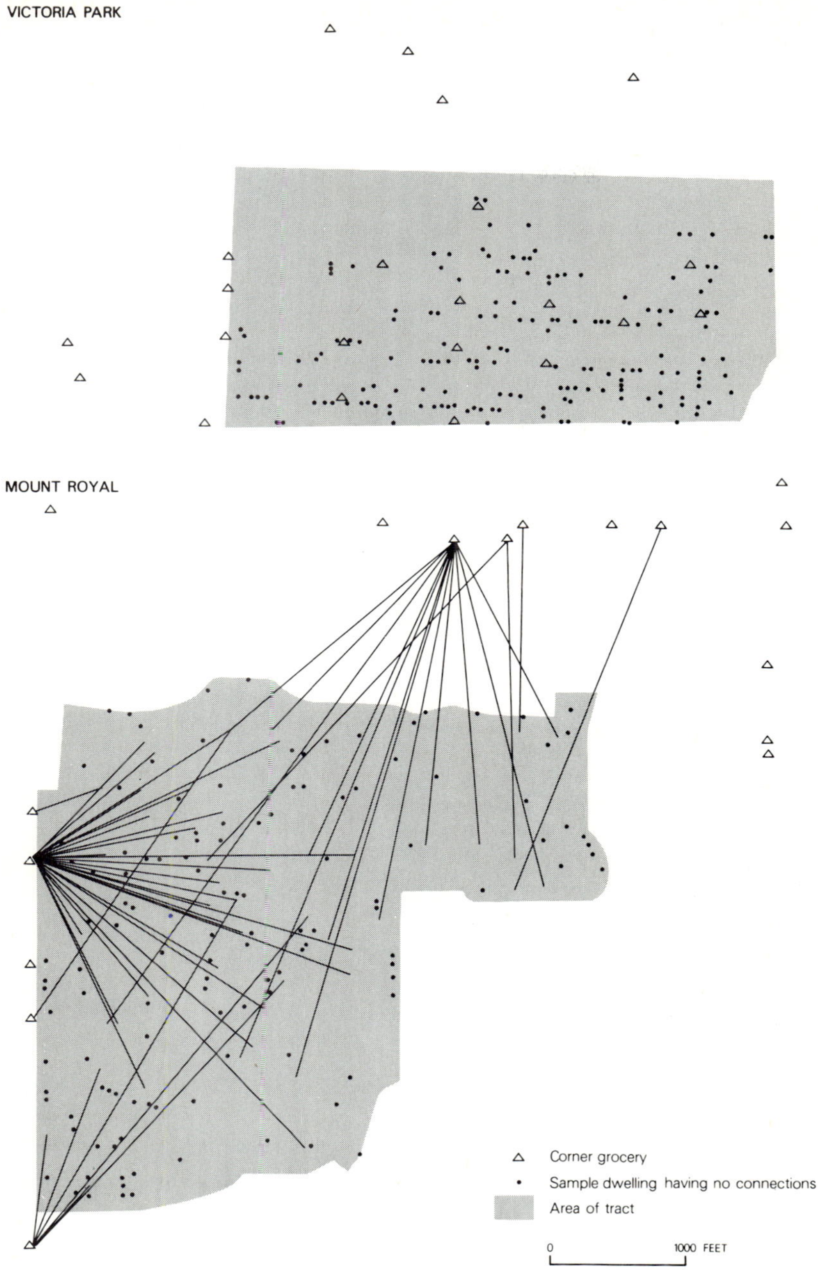

FIGURE 9,6 Car linkages - H.D.I.A. - corner grocery shops.

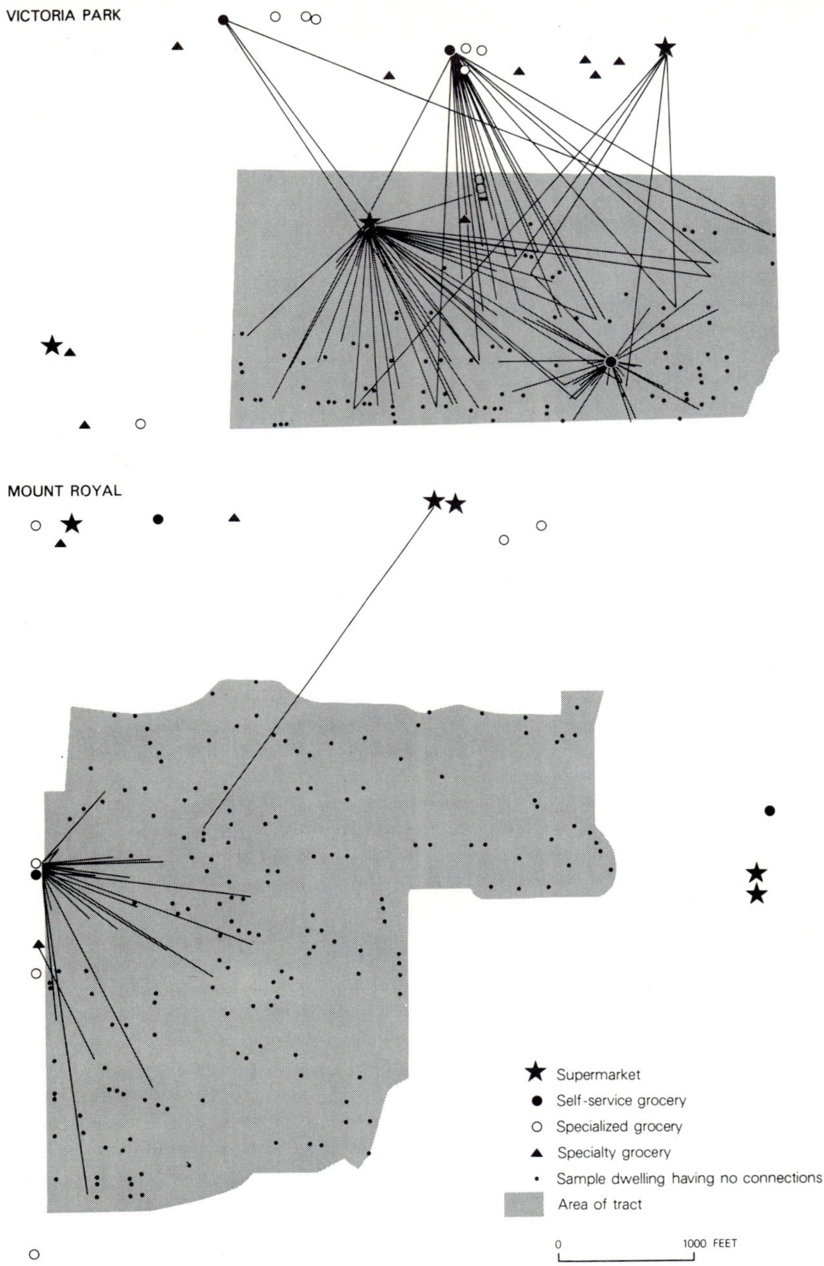

FIGURE 10,6 Pedestrian linkages - H.D.I.A. (excluding corner grocery shops)

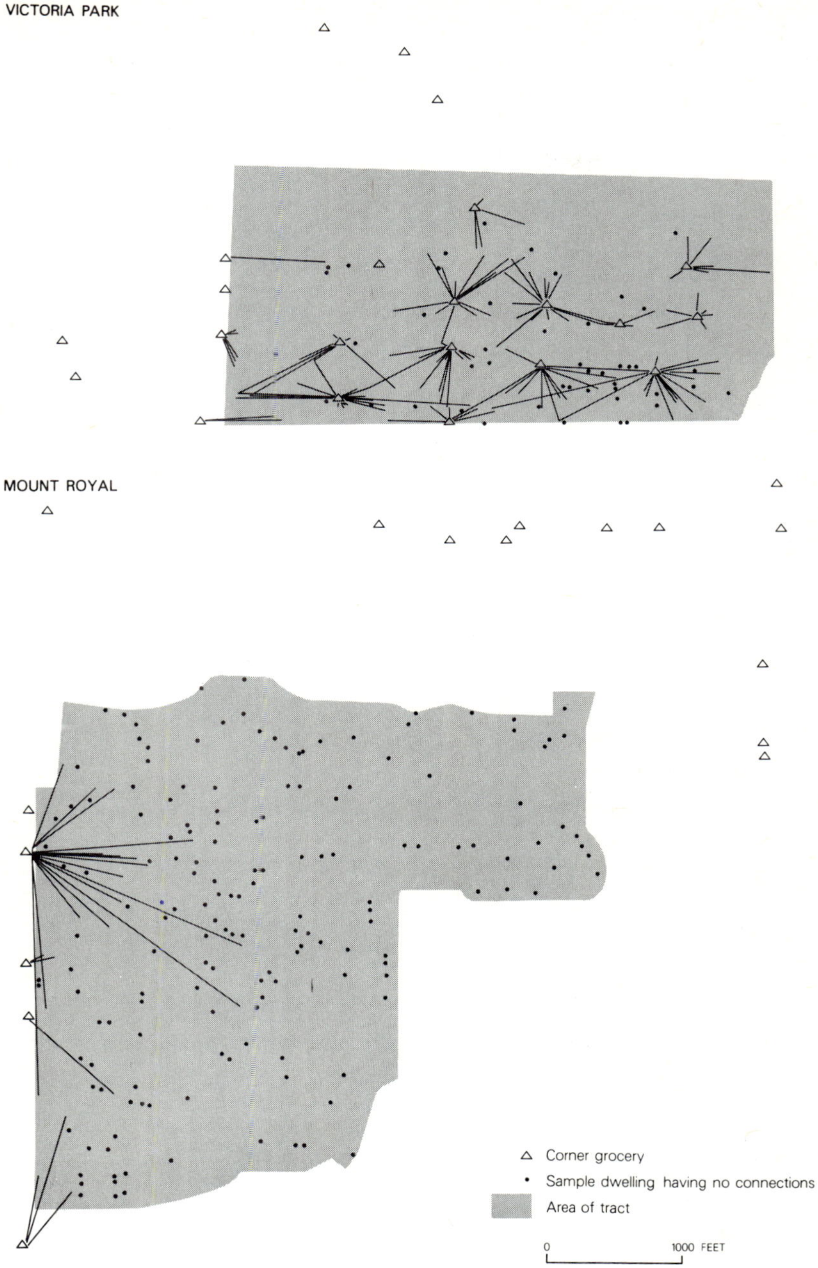

FIGURE 11,6 Pedestrian linkages - H.D.I.A. - corner grocery shops.

roof. By patronizing groups of closely spaced specialized stores (food centres) this need can also be satisfied.

Individual Stores Frequented

Of the 147 food stores mapped in the sector, forty-six were enumerated by at least one Mount Royal respondent, while twenty-nine were listed by Victoria Park respondents (Figure 12,6). The most popular store for the Victoria Park respondents was the Co-Op supermarket in the Victoria Park tract (listed by 52 per cent of the sample). Generally, Victoria Park dwellings interacted with a supermarket, self-service, and corner grocery stores within the tract, and downtown areas (Figures 12,6 and 13,6). Mount Royal respondents interacted with a greater variety of store types and locations, the Woodwards supermarket in the Chinook regional shopping centre being the most frequented (by 47 per cent of the sample). A self-service grocery store, specialized and corner grocery stores on the western boundary of Mount Royal were patronized by at least ten per cent of the sample, with Fairley's self-service being the store most frequented in this area (Figures 12,6 and 13,6).

Pairs of Stores Frequented

Examination of pairs of stores by one or more sample dwellings reveals that respondents in both tracts, especially in Mount Royal, interacted with a variety of pairs of food stores. One hundred fourteen different pairs of stores were connected with one or more respondents in Victoria Park. Fifty-six of these pairs were connected with only one respondent (Table 7,6). Mount Royal respondents listed 372 different pairs; 247 of these were enumerated by only one respondent.

Differences between tracts are evident when the pairs of stores are classified according to store type (Table 8,6). Combinations of a supermarket and a corner grocery store account for almost fifty per cent

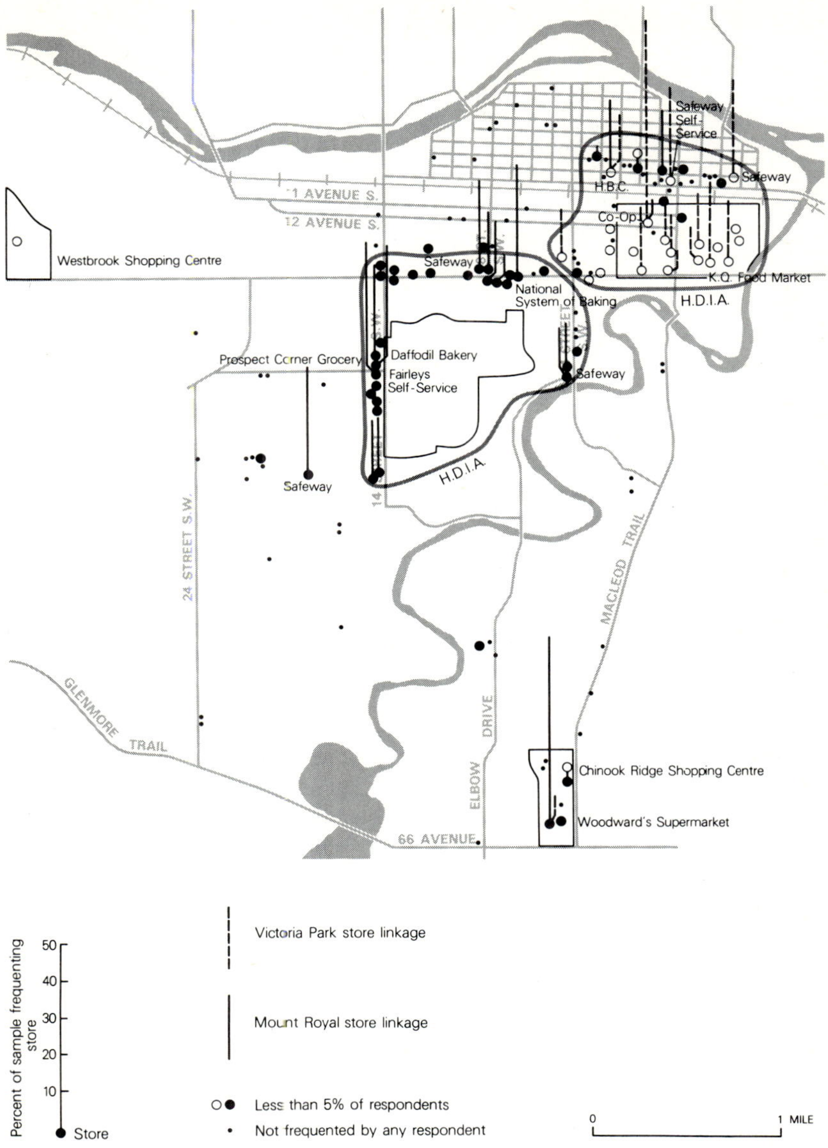

FIGURE 12,6 Victoria Park and Mount Royal linkages with individual stores.

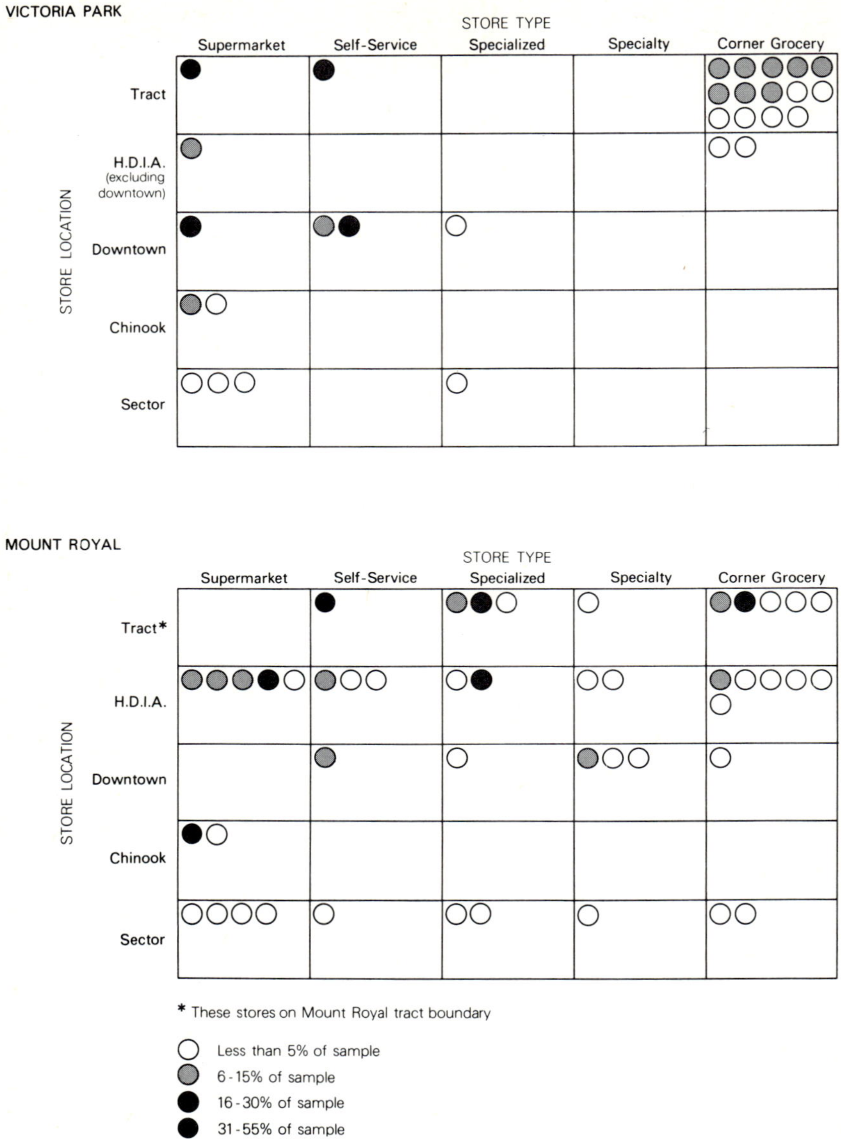

FIGURE 13,6 Types and locations of individual stores frequented by Victoria Park and Mount Royal respondents.

of the Victoria Park pairs. The Mount Royal tract was not dominated by any one pair of store types. The leading pair was the supermarket-specialized food store combination.

TABLE 7,6

COMBINATIONS OF TWO AND THREE FOOD STORES FREQUENTED BY SAMPLE RESPONDENTS

Number of Respondents Linked with Identical Combinations of Stores	Number of Combinations			
	Two Stores		Three Stores	
	Victoria Park	Mount Royal	Victoria Park	Mount Royal
1	56	247	81	515
2	20	51	14	80
3	11	25	3	22
4	7	11	1	14
5	6	11	0	11
6	4	5	0	4
7	3	6	0	2
8	2	4	0	1
9	0	5	0	1
10	0	1	0	0
11	1	3	0	5
12	2	1	0	0
13	1	2	0	0
15	1	0	1	0
16	0	0	1	0
17	0	0	2	0
18	0	0	1	0
24	0	0	1	0
25	0	0	1	0
Total	114	372	105	645

Source: Interview data, 1968.

Few pairs of stores were distributed at the intra-centre scale, in fact, only one pair for Victoria Park was recorded (Table 8,6). This result casts some doubt on the validity of the food centre concept as defined in this study, particularly in the case of the Mount Royal responses where food shopping was done at a variety of pairs of food stores throughout the sector. The food centre concept is more valid for Victoria Park, simply because respondents in this tract frequent few stores (Table 4,6). Pairs consisting of supermarket and a corner grocery,

TABLE 8,6

PAIRS OF STORES FREQUENTED ACCORDING TO STORE TYPE

Pairs of Stores	Victoria Park		Mount Royal	
	Number	Percent	Number	Percent
Supermarket-Supermarket	45	15	101	10
Supermarket-Self Service	36	12	107	11
Supermarket-Corner Grocery	146	49	130	13
Supermarket-Specialized	6	2	263	26
Supermarket-Specialty	0	0	18	2
Self-Service-Self Service	7	2	8	1
Self-Service-Corner Grocery	39	13	40	4
Self-Service Specialized	0	0	72	7
Self-Service-Specialty	0	0	15	1
Corner Grocery-Corner Grocery	11	4	13	1
Corner Grocery-Specialized	0	0	97	9
Corner Grocery-Specialty	0	0	7	1
Specialized-Specialized	0	0	41	4
Specialized-Specialty	0	0	18	2
Specialty-Specialty	0	0	1	-
Out of Sector	8	3	85	8
Total	298	100	1016	100
Intra-Centre Pairs	1	-	65	6

Source: Interview data, 1968.

PLATE 24
The new Mount Royal College campus, on a former military base, southwestern Calgary.

PLATE 25
The Southern Alberta Institute of Technology.

City of Calgary Photo

or a self-service grocery and a corner grocery account for sixty-one per cent of the Victoria Park pairs (Table 8,6). It appears that Victoria Park respondents use these one-store food centres for most of their needs, and rely on corner grocery stores (non-food centres) for occasional food purchases. Mount Royal respondents appear to shop at scattered locations within the sector, and pairs of stores account for only a part of their total shopping needs.

Triads of Stores Frequented

Although a great number of combinations of three stores were listed, a small proportion of these were linked with more than one respondent (Table 7,6). The smaller number of triads in Victoria Park was due mainly to the fact that only thirty per cent of the respondents were linked with three or more stores (Table 4,6). Seventy-one per cent of the Mount Royal dwellings were linked with three or more stores, and 645 triads were associated with respondents in this tract. Of all the triads in both tracts, only two were intra-centre. A combination of a self-service (Fairley's), bakery (Daffodil), and corner grocery (Prospect), were listed by eleven respondents, and a triad of meat market and two corner grocery stores was listed by two respondents. Both of these triads were cited by Mount Royal respondents.

CONCLUSION

The linkage concept facilitated comparison of the reaction of sample respondents in the two study tracts to the forms and patterns of food stores in the southern sector of Calgary. Although food store-dwelling linkages for both tracts were restricted mainly to the study sector, the nature of these linkages was different. These differences are summarized by comparing the three components of the food store-dwelling linkage, the dwelling, the connection, and the stores, and the store combinations linked with the sample dwellings.

The analysis of the linkage components revealed that Mount Royal food shopping patterns were more complex than those of Victoria Park. This complexity can be partially explained by the greater mobility and food storage potential of dwellings in the Mount Royal tract. Automobile connections were dominant for the Mount Royal respondents, even for relatively short distances. Victoria Park respondents relied mainly on the pedestrian mode. The complexity of Mount Royal linkages was reflected by the greater number of linkages per dwelling. Food stores frequented by Victoria Park respondents were restricted mainly to the Victoria tract and its high density interaction area, including the nearby downtown area. Mount Royal dwellings were linked with a variety of store types and locations throughout the study sector.

The examination of store combinations linked with the sample dwellings reveals that the two tracts differed in terms of how their residents interacted with food stores in the sector. Victoria Park respondents interacted with fewer stores than those in Mount Royal and this interaction was restricted mainly with supermarkets, self-service, and corner grocery stores in the tract and H. D. I. A. Mount Royal responses indicated that food shopping occured throughout the sector, and this shopping involved a great variety of store types. If one-store food centres are ommitted, the food centre concept has little application, particularly in the Mount Royal situation. Not only were intra-centre pairs or triads few in number, but in the case of Mount Royal, two or more stores were seldom associated with more than one respondent. As a general comparison, Victoria Park respondents reflected their lower income and mobility by interacting with local stores, particularly with supermarkets, self-sevice, and corner grocery stores. Mount Royal respondents reflected their higher income and mobility by choosing a great variety of stores in terms of store location and store type. Respondents in both tracts took advantage of the convenience of one or more supermarkets, but Mount Royal respondents interacted with a number of supermarkets scattered throughout the sector, and complemented these linkages by frequenting self-service, specialized, and specialty stores.

This study was primarly descriptive. However, some observations regarding the urban retail structure can be made. The concept of the shopping centre, implying interaction between dwellings and contiguous groups of stores was not supported by the interview data. This raises questions about traditional interpretations of retail structure which treat the "centre" as an important component, particularly when central place theory is applied at the intra-urban scale. It can be hypothesized that consumers treat centres and other isolated stores as dispersed; shoppers visit a variety of separated centres and stores on one or more shopping trips although, theoretically, one centre could provide all of their needs. This study was limited, however, in terms of the types of residential areas considered. Further studies, involving all types of shopping, and analysis of actual shopping trips, are needed. These studies would have to consider a better representation of residential areas in the city, particularly those in the middle class category. Whether more extensive studies will support this hypothesis remains to be seen.

REFERENCES

1. MITCHELL, R.B. and RAPKIN, C. Urban Traffic: A Function of Land Use. New York: Columbia University Press (1954), p. 111.

2. DOODY, A.F. and DAVIDSON, W.R. "Next Revolution in Retailing," Harvard Business Review, 45 (1967), p. 5.

3. LASKI, Marshannta. The Gardian. Manchester (n.d.).

4. LEWIS, D., Editor. The Pedestrian in the City. Princeton: D. Van Nostrand Co., Inc. (1966).

5. NYSTVEN, J.D. Geographical Analysis of Customer Movement and Retail Business Locations. Unpublished Ph.D. dissertation, University of Washington (1959), p. 78.

6. BOAL, F.W. An Analysis of Retail Business Location and Customer Movement in Ann Arbor, Michigan. Unpublished Ph.D. dissertation, University of Michigan (1963), p. 33.

7. GARRISON, W.L. et. al. Studies of Highway Development and Geographic Change. Seattle: University of Washington Press (1959), p. 69.

8. BERRY, B.J.L. Commercial Structure and Commercial Blight. University of Chicago, Department of Geography Research Paper 83 (1963); SIMMONS, J. The Changing Pattern of Retail Location. University of Chicago, Department of Geography Research Paper 92 (1964); and GARNER, B.J. The Internal Structure of Retail Nucleations. Northwestern University, Studies in Geography, 12 (1964).

9. SIMMONS, J. Toronto's Changing Retail Complex, University of Chicago, Department of Geography Research Paper 104 (1966).

10. JONES, E. A Social Geography of Belfast. London: Oxford University Press (1960), pp. 85-89.

11. BOAL, F.W., op. cit., p. 53; and JOHNSON, D.B. A Functional Comparison of the Central Retail District With Two Regional Shopping Centres in Clagary, Alberta, Unpublished M.A. thesis, University of Alberta (1963), p.95.

12. SIMMONS, J. (1964), *op. cit.*, p. 2.

13. *Census of Canada, 1968.* Census Tract Bulletin - Calgary, Ottawa: The Queen's Printer (1968), p. 8.

14. ARKIN, H. and COLTON, R.R. *Tables for Statisticians.* New York: Barns and Noble, Inc. (1950), p. 136.

15. Note: this section involves the analysis and discussion of linkages involving movement of the consumer. Home delivery and other modes are excluded (Table 7,6).

PLATE 26
Calgary Exhibition and Stampede,
Victoria Park residential area.

Calgary Exhibition and Stampede Photo

PLATE 27
Calgary Exhibition and Stampede in the 1950's; note the skyline prior to the post-1960 office building boom.

Alberta Government Photo

CHAPTER 7

A MULTIVARIATE DESCRIPTION OF CALGARY'S COMMUNITY AREAS

Wayne K.D. Davies

University of Calgary

INTRODUCTION

The objective of this chapter is to demonstrate the utility of multivariate techniques in delineating the social and demographic character of the communities of Calgary using 1971 census data (Figure 1,7). Although it may be trite to repeat the point, it is nevertheless true that cities contain the most complex variety of human groups. These groups interact and overlap in a spatial context, producing a complex social and economic mosaic. The resultant heterogeneity is often considered to be one of the hallmarks of urbanism[1] and is as much a feature of a youthful city such as Calgary, which has grown from a police fort to a centre of over 400,000 in a mere one hundred years, as it is of Jerusalem, Rome, or London, although obviously the age, size, and greater cultural mixture of these three world cities produce even greater degrees of differentiation. But despite differences in the range of intraurban variation, the problem facing any investigator attempting to capture and portray the social patterns and character of urban communities is the same in all cases; it is the choice of the descriptive form or medium to be adopted. So before dealing with the specific case of Calgary it is necessary to outline the variety of alternative approaches that can be used.

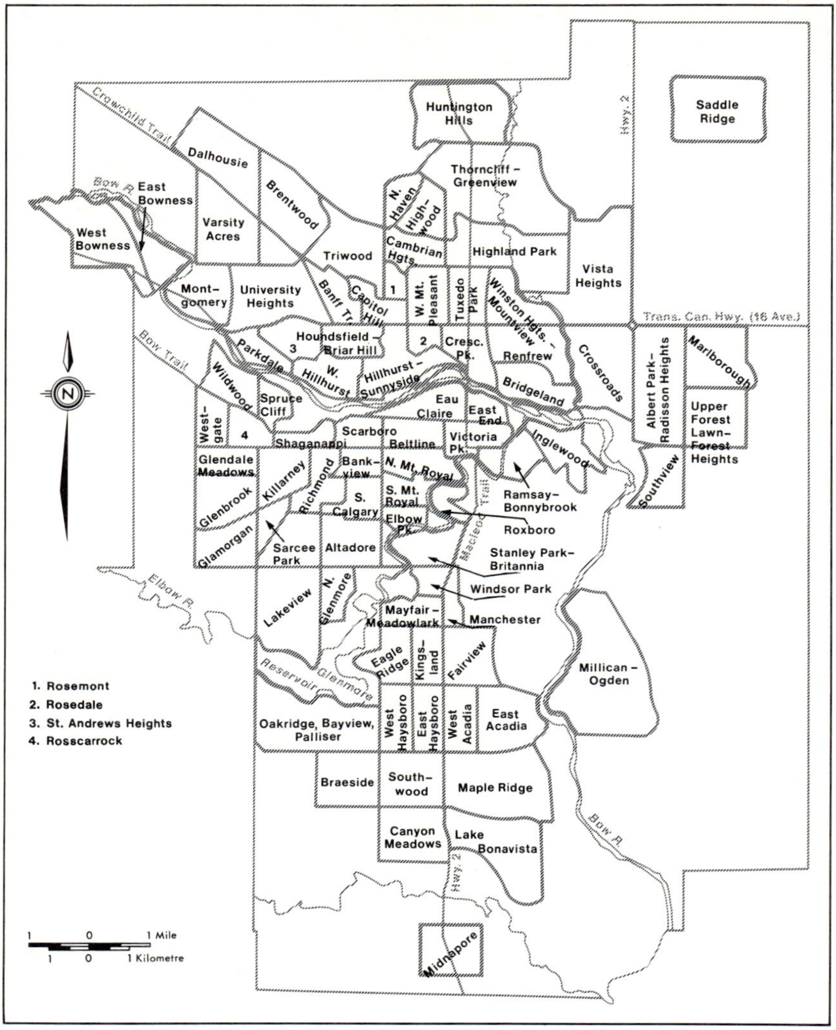

FIGURE 1,7 Location of the Calgary Communities, 1971.

ALTERNATIVE APPROACHES TO DESCRIPTION

The study of the character and differential nature of intra-urban areas, or more specifically, communities, can be divided into four developmental phases and two distinctive purposes, elements of all of which can be found in the contemporary descriptions of urban areas. The first phase, the 'Impressionistic" is represented by the descriptive vignettes of urban areas produced by novelists as scenes or background to their story. Whether written in the classic mould of Charles Dickens, or in the action orientated lighter vein of twentieth century urban crime writers such as Raymond Chandler or Ed McBain, the by-product can be an illuminating characterization of urban places and conditions. The second phase, called 'Investigative Reporting', is more specific, despite being primarily impressionistic in nature. Here the characterization of urban life and places is the focal point of the description, rather than being an incidental part of another purpose. This approach has been used by a long line of social commentators beginning with Mayhew[2] and Engels[3] in nineteenth century England, or Woodsworth[4] in Canada. The third phase, that of 'Statistical Description', emerged only gradually from the investigative approach. Although the objective of the description is the same, precise factual summaries replace the qualitative verbal discussion. In extreme form this is the approach of the national census authorities, although here the collection of figures becomes the end in itself. But where interpretation and description is a primary objective the work of Charles Booth[5] in late nineteenth century London is of vital historical significance. Booth discovered that information on the distribution of poverty did not exist in accessible form. So he set about collecting, and more to the point, interpreting, data on the poverty in the city using precise quantitative figures. Finally, a fourth phase can be recognized, that of 'Statistical Analysis'. The crux of the approach consists of manipulating the type of statistical information used in the previous phase to build a series of indices, or profiles, that summarized urban patterns and conditions. These indices provide measurements of the dimensions of variation existing in the variable set, and from these it is possible to describe the differences between communities in the cities. This approach, represented by the application of Factor Analysis, is the one that will be used in this chapter. Yet we must remember that factorial techniques are nothing more than a sophisticated version of the heritage that began with Booth. He averaged six indicator variables of poverty and low social status to obtain a summary index of the social conditions in London and this formed the basis of his Presidential Address to the Royal Statistical Society in 1893.[6] Clearly this was similar in technique to the social area indices formulated by Shevky and Bell[7] in 1955, although in this latter case standardization procedures were used, and more than one axis of differentiation was identified.

Recognition of a series of developmental phases does not, however, complete the range of alternative approaches. It fails to answer the question of what the description is used for. To novelists, the purpose of the description is clear, it is background. For all the others two approaches are possible. First of all it can simply be used empirically, as a simple verbal or factual statement on urban conditions or patterns. Secondly, it can be used to tease out the principles of variation within cities, or to provide concepts that can be used as building blocks for the development of an interrelated body of knowledge on urban social variation. Formalized as an academic approach the creation of a systematic study of ecological character of cities was the achievement of the so-called 'Chicago school' of human ecology, represented by Park,[8] Burgess and McKenzie, although we must recognize that students of many other disciplines were carrying out parallel work in allied fields such as Geography, Sociology, Economics, History, as well as Planning.

In the contemporary literature on urban description, examples of all four developmental phases can still be found. In addition, there are idiographic city studies as well as those used to derive basic principles of urban social variation that can contribute to a body of theoretical literature on urban character. Yet within the confines of a single study it is impossible to adopt all of these approaches simultaneously. A choice must be made, and fundamentally this is related to the values of the investigator as well as to his purpose. In this study the values are those of making the description as precise as possible. In other words the description is placed upon a quantitative basis to ensure that the generalizations produced are mathematically, rather than intuitively derived, thereby ensuring that they are capable of replication by other workers. To achieve this end, factor and cluster analysis techniques from the Statistical Analysis phase are used. Factor analysis produces a parsimonious description of the sources or dimensions of variation in a data set by showing how variables are associated together, as well as producing measurements of the contribution of the individual observation units to those clusters of variation via factor scores.[9] Cluster analysis can then be applied to the scores to summarize the regional patterns.[10] To avoid misunderstanding it must be stressed that the adoption of this approach, like all the others, contains strengths and weaknesses. These issues are more properly dealt with in the conclusion to this chapter. However, given the need for precision, and the desire to build up a body of generalizations on the dimensions of urban social variation that may ultimately be testable in a scientific sense, the adoption of a multivariate framework seems inevitable, with preuse description as the first objective.

DATA AND TECHNIQUES

Factor analysis methods can be used in either a hypothesis-testing or an exploratory context by using the common factor or component models respectively. Given the objectives of this study the R mode component model was considered to be the most appropriate approach. However it had two other advantages, namely that the calculated component scores did not have to be estimated, whilst assumptions did not have to be made either about the number of common factors or the communalities.[11] The Principal Axes factoring technique was applied to a data set consisting of sixty variables (Appendix 1). Groups of from three to six variables were used to measure the following nine basic sets of variation: age (5); family (5); female characteristics (4); dwelling type and tenure (6); households (3); work participation and education (6); occupation and income (6); migration (6); age of construction (6). This type of equivalence between the categories could not, however, be maintained for the ethnicity variables, and the range of variation here could only be accommodated by using fourteen variables. In the calculation of those variables attempts were made to avoid problems of overlap and spurious correlations.[12]

Using the 'rule of thumb' methods proposed by Davies[13] and Barrow, a series of alternative solutions were evaluated in which various rotations were applied and different numbers of axes were abstracted from the 60 x 60 Pearson Product Moment Correlation Matrix between the variables.[14] The most stable and interpretable result was a six axis solution using the Direct Oblimin oblique rotation with gamma set at 0.0. This solution accounted for 67.9 per cent of the initial variance. The degree of parsimony in this result can be judged from the fact that the new 60 variable x 6 axes factor loading matrix accounts for over two thirds of the original variation in the data set.

THE STRUCTURAL AXES AND DISTRIBUTION PATTERNS

Table 1,7 shows the loadings greater than $\pm$ 0.3 associated with each of the six axes of differentiation produced by the factor analysis. In every case a simple descriptive title has been given to summarize the pattern of loadings. The largest axis, accounting for 17.4 per cent of the overall explanation, forms a clearly defined scale. On the positive side are a set of variables indexing family characteristics, namely high fertility levels, single detached owner occupied dwellings and high percentages of children; on the negative side are the non-family characteristics, of which young adults, apartment dwellings, lodgers, single persons and high participation levels have the highest loadings. This axis conforms to the familiar Family Status dimension recognized by many social area and factor analysts, from Shevky and Bell[15] in San Francisco to Murdie[16] in Toronto. In passing it is worth noting that the designation 'family status' is more suitable for Calgary than the 'life cycle' term applied to a polarization between young and old people. Perhaps this provides additional evidence for Timms's[17] point that family status becomes a more recognizable axis than life cycle in the larger or more modern cities.

The distribution of component scores on the first dimension (Figure 2,7) shows the broadly concentric zonation first described by Engels[18] but formalized by Burgess.[19] Non-family characteristics are found in the recently developed apartment complex on the west and south of the commercial center, where Eau Claire, Belt Line and North Mount Royal all have scores over +3.00. A systematic decline in the size of these values occurs as one moves outwards from the core, although the pattern is truncated in the relatively poorer eastern communities. The result is a gradual transition to high negative values indexing high family characteristics in the most recently developed suburbs, where Lake Bonavista, Braeside, Varsity Acres and Marlborough all show values over -1.00. Only three exceptions to this concentric zonation occur. One is pivoted on the junction of the Glenmore and Macleod Trails in the South West, and is accounted for by the developing apartment communities in parts of Windsor Park and Kingsland. A less intense area of non-family

TABLE 1,7

LOADINGS OF VARIABLES ON THE 6 COMPONENT SOLUTION: CALGARY, 1971

Component 1	FAMILY STATUS (17.4%)
Loadings	Short Title of Variable
+0.76	12) Fertility Ratio
+0.73	15) Owner Occupied
+0.71	16) Single Detached Dwelling
+0.62	8) Family Size
+0.61	9) Number of Children
+0.59	1) Children
+0.58	21) Persons per Household
+0.41	23) Large Households
+0.40	3) Mature Adults
+0.37	35) Household Income
+0.35	20) Large Dwellings
-0.32	49) Native Ethnic
-0.40	11) Females
-0.40	41) Provincial Migrants
-0.53	7) Divorced
-0.54	24) Participation Ratio
-0.59	22) Small Households
-0.61	38) High Intermunicipal Movers
-0.63	19) Small Dwellings
-0.64	6) Single Adults
-0.67	25) Young Adult Participation
-0.70	13) Female Heads
-0.71	10) Lodgers
-0.77	17) Apartments
-0.85	2) Young Adults

Component 2	SOCIO-ECONOMIC STATUS (14.3%)
Loadings	Short Title of Variable
+0.88	33) Manufacture
+0.87	34) Construction & Transportation
+0.69	46) German Ethnic
+0.62	27) Limited Education
+0.45	53) Italian Ethnic
+0.37	55) Polish Ethnic
+0.35	51) Netherlands Ethnic
+0.34	45) French Ethnic
+0.32	57) Dwellings built 1946-50
-0.32	41) Provincial Migrants
-0.40	6) Single Adults
-0.64	44) British Ethnic
-0.67	50) Jewish Ethnic
-0.72	35) Household Income
-0.75	32) Vocational
-0.84	20) Large Dwellings
-0.84	31) Managerial
-0.85	29) Graduates
-0.89	30) White Collar Ratio

Component 3	MIGRANT STATUS (11.4%)
Loadings	Short Title of Variable
+0.93	36) Non-migrants
+0.91	37) Low Intermunicipal
+0.81	4) Middle Aged
+0.75	58) Dwellings built 1951-60
+0.54	43) Canadian Citizens
+0.36	6) Single Adults
-0.54	41) Provincial Migrants
-0.61	3) Mature Adults
-0.64	38) High Intermunicipal movers
-0.80	40) Recent Migrants
-0.88	60) Dwellings built 1966-70

TABLE 1,7 (continued)

LOADINGS OF VARIABLES ON THE 6 COMPONENT SOLUTION: CALGARY, 1971

Component 4	ECONOMIC PARTICIPATION - NON AFFLUENT (11.3%)	Component 5	CANADIAN-ETHNICITY (6.2%)	Component 6	FAMILY SIZE (EAST EUROPEAN) - OLD AGE (7.3%)
Loadings	Short Title of Variable	Loadings	Short Title of Variable	Loadings	Short Title of Variable
+0.76	11) Females	+0.81	42) Canadian Stock	+0.60	47) Ukrainian Ethnic
+0.54	28) High School Education	+0.58	43) Canadian Citizen	+0.58	59) Dwellings built 1961-65
+0.49	24) Participation Ratio	+0.51	45) French Ethnic	+0.50	23) Large Households
+0.42	2) Young Adults	+0.38	44) British Ethnic	+0.47	9) Number of Children
		+0.36	28) High School Education	+0.46	8) Family Size
-0.30	22) Small Households	+0.32	24) Participation Ratio	+0.40	21) Persons per Household
-0.30	54) Hungarian Ethnic	+0.32	51) Netherlands Ethnic	+0.36	54) Hungarian Ethnic
-0.33	19) Small Dwellings	+0.30	52) Scandinavian Ethnic	+0.36	55) Polish Ethnic
-0.38	45) French Ethnic			+0.35	1) Children
-0.44	6) Single Adults	-0.50	39) Immigrants	+0.35	3) Mature Adults
-0.47	5) Old Aged	-0.68	53) Italian Ethnic	+0.33	24) Participation Ratio
-0.47	39) Immigrants			+0.31	46) German Ethnic
-0.51	7) Divorced				
-0.53	27) Limited Education			-0.35	13) Female Heads
-0.58	49) Native Ethnic			-0.38	22) Small Households
-0.67	48) Asiatic Ethnic			-0.38	44) British Ethnic
-0.84	26) Unemployed			-0.46	56) Dwelling built pre 1946
-0.93	18) Substandard			-0.50	5) Old Aged
				-0.67	57) Dwelling built 1946-50

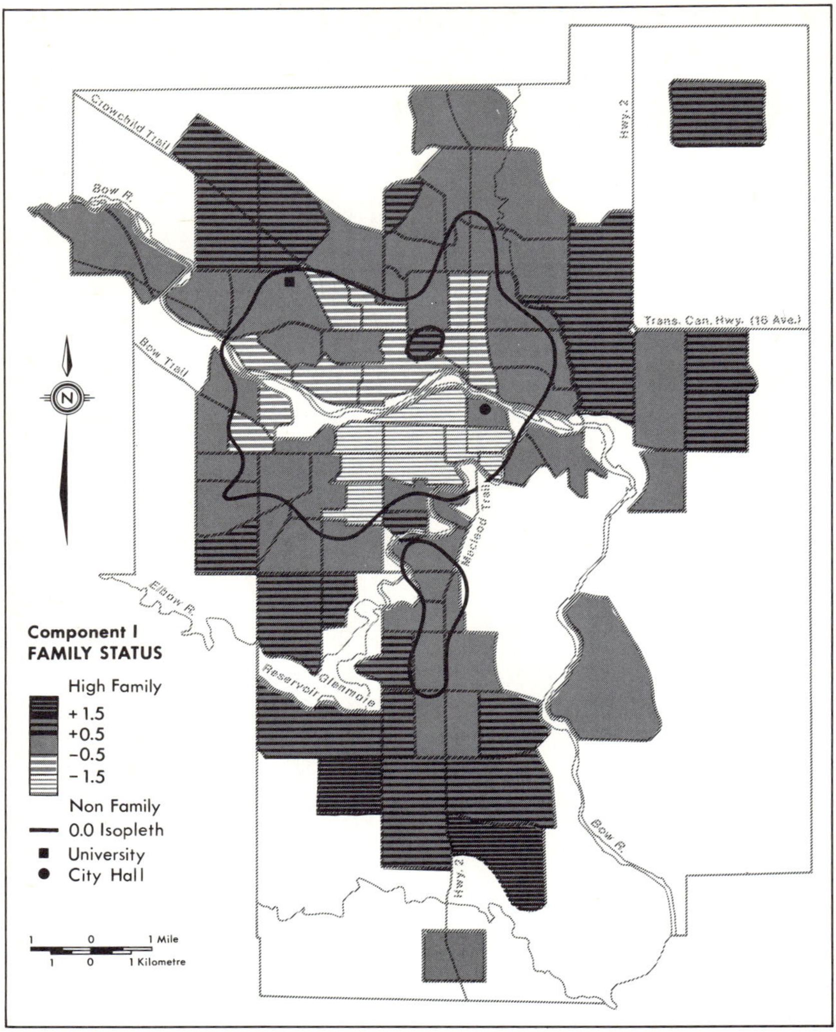

FIGURE 2,7 Distribution of scores on the family status component.

characteristics is found around the main commercial strip north of the city centre, beginning near 16th Avenue and Center Street and spreading westwards into Hillhurst. Similarly the apartment complex of Spruce Cliff and public housing projects ir Rosscarrock both show high non-family characteristics.

The second largest axis explains 14.3 per cent of the total variance and is the equally familiar dimension of Socio-Economic Status. The negative side of the vector is the one indexing high status characteristics, in particular, the white collar ratio, percentage of graduates, people in managerial occupations and dwellings with over 8 rooms. Significantly, the two ethnic groups loading most strongly on this axis are those of British origin and the Jewish religion. By contrast, the positive loadings pick out lower status characteristics, such as percentage of employed people in manufacturing and construction industries and those of limited formal education. Many western European ethnic groups are also associated with this side of the vector in addition to Poles and Italians, demonstrating that ethnicity is clearly linked to socio-economic status in the city at least as far as ecological correlations are concerned. Spatially, the pattern of scores on the socio-economic dimension shows a completely different pattern to the family status map (Figure 3,7) . The whole eastern side of the city is of low status, with a clear north-south division running from Thorncliffe through West Mount Pleasant to Eau Claire, Roxboro, Windsor Park, along Macleod Trail to East Haysboro and Southwood, and then eastwards along the Acadia-Maple Ridge boundary. To the west are the high status areas. The highest values occur in a southeast trending sector running from South Mcunt Royal and Elbow Park through to Eagle Ridge, with an outlier in the University and St. Andrews Heights area of the northwest. The only exception to this east-west division of the city is found in the Bowness-Montgomery salient of low status characteristics which occupies the Bow Valley flood plain, and to lesser extent in the

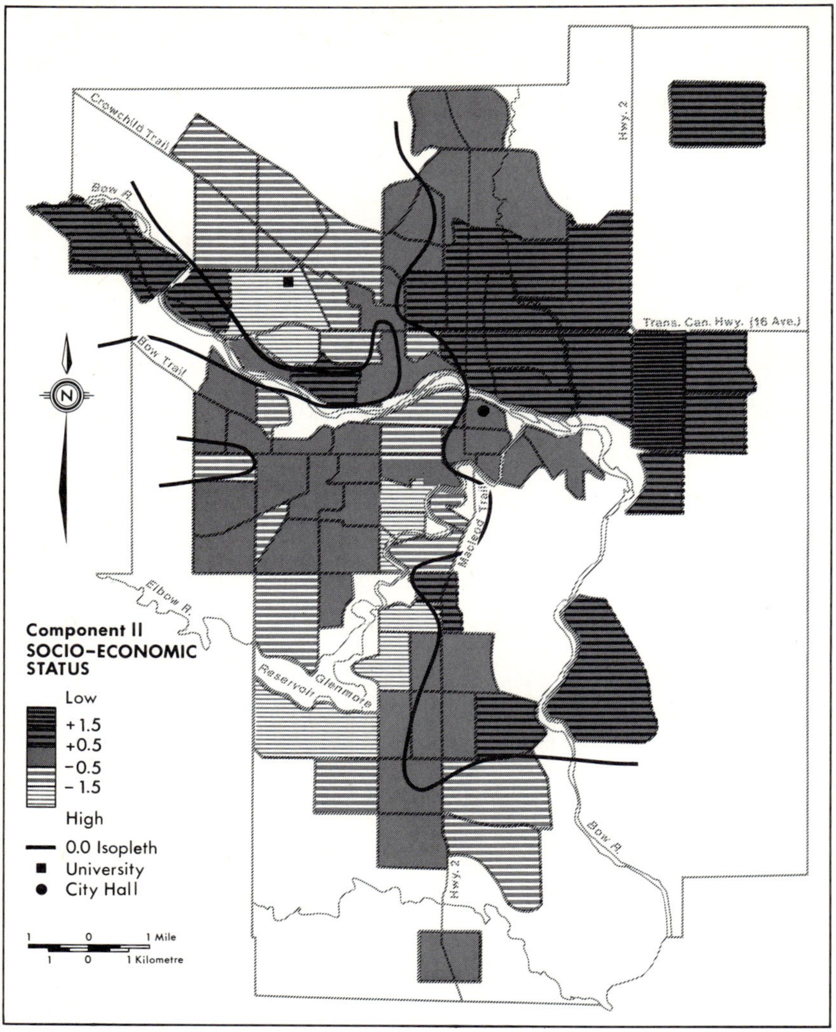

FIGURE 3,7 Distribution of scores on the socio-economic status component.

Westgate, Glenbrook, Killarney, South Calgary, Bankview sector running southwards from 17th Avenue S.W.

The third axis of differentiation identified in the study distinguishes non migrant from high migrant characteristics and accounts for 11.4 per cent of the overall variance. It is called Migrant Status. The positive loadings are dominated by non-migrants, people who have not moved houses or changed their city of residence in the previous five years, as well as by people in the 45 to 65 year old categories, and by dwellings constructed in the 1950's. Negative loadings are associated with : dwellings built in the late 1960's; people who have moved in the past year; with high levels of movement between cities; and with persons in the 25 - 44 year old group. Comparatively few factorial studies of urban character have identified this axis, although it is a vital part of the 'complexity of society' construct postulated by Shevky and Bell. But even these pioneers of social area studies failed to provide a separate index to identify the axis. In distributional terms two areas of the city display high migrant characteristics (Figure 4,7). First of all are the most recently developed suburbs on the outer fringes of the city, with extensions to the university area on the North West, and along Macleod Trail to Kingsland in the south. The second zone comprises the inner city area from Hillhurst to South Calgary and Victoria Park, with an outlier in the Spruce Cliff community. We have already seen that the two zones are quite different in family characteristics. Hence it is apparent that high mobility levels are associated with the new family orientated suburbs as well as with the apartment and rooming house areas of the central city where young adults and post family groups are to be found.

Although the socio-economic axis provided a scale of social status in the city, it is significant that the variables identifying the most impoverished groups did not load on the dimension. Instead they are found on the fourth axis of the study called Participation - Impoverishment which

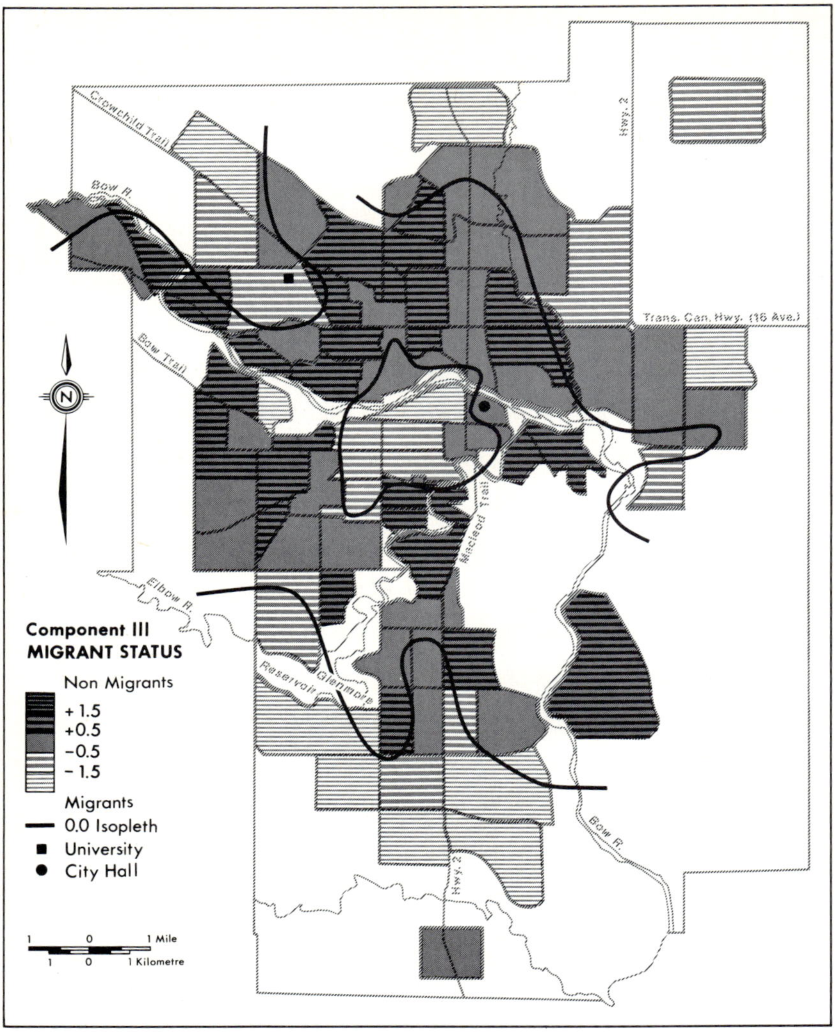

FIGURE 4,7 Distribution of scores on the migrant status component.

explains 11.3 per cent of the variance. This axis does not form a simple scale in the same way as the other axes. Instead it is bi-polar. The negative loadings identify the impoverished or non-affluent groups with characteristics such as sub-standard houses, unemployed, Asiatic and Native Ethnic groups, people of limited education, recent immigrants and old aged. The positive loadings, on the other hand, pick out high concentrations of females, those with high school education, high participation levels and young adults. The map of the distribution of scores on the axis (Figure 5,7) is dominated by an eastern and inner city sector of high negative loadings. Beginning in the South Calgary, Hillhurst and Tuxedo Park communities, and extending south eastwards through to Millican and Ogden, the core of the area is in the East End, Victoria Park, Inglewood and Ramsay communities. This area, together with outliers in the Albert Park - Forest Lawn and East Bowness-Montgomery communities, is the zone where the least affluent groups are located. Significantly this is on the eastern edge of the inner city, the area often called the 'zone of discard' as for example by Murphy and Vance[20], and merges into the industrial eastern side of the city. The high positive scores, the areas of high participation, are in the older family suburbs already designated as being of middle or low economic status. Their principal location is in a sector running from the Wildwood, Parkdale, Briar Hall and Cambrian Heights areas, through to the newer communities of Huntington Hills and Highland Park. This type of 'broken collar' distribution in the inner suburbs shows the same type of location for high participation levels found by Davies and Lewis[12] in Leicester, England.

The fifth component, labelled Canadian and Ethnicity, is basically one measuring certain types of ethnicity. The positive side of the axis is principally associated with Canadian citizens and those with two parents that are Canadian born. To a lesser extent it is linked with people of North West European ethnicity, high levels of participation, and high-

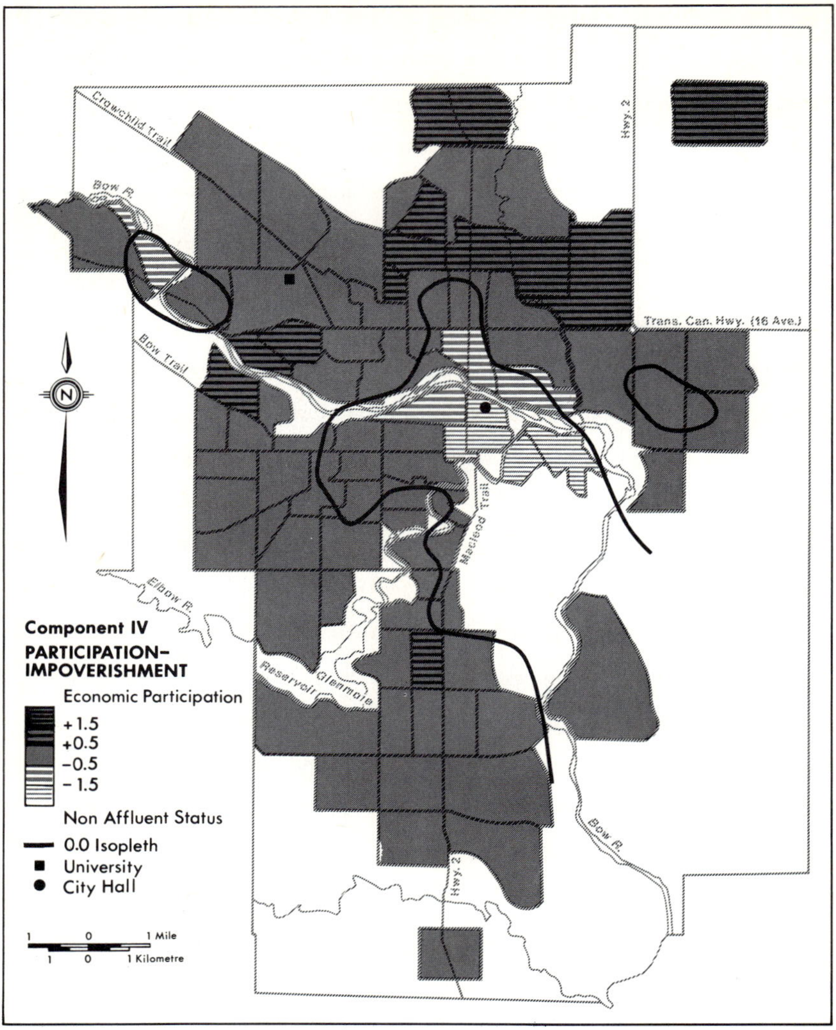

FIGURE 5,7 Distribution of scores on the participation-non affluent component.

school education. Inversely related to these characteristics are immigrants and people of Italian stock. The negative scores associated with these factors form the primary feature of the distribution patterns (Figure 6,7). In essence the highest scores form a crescent in the inner city, from Killarney through University Heights, Rosemont, Thorncliffe to Ramsay and Hillhurst, with outliers in Dalhousie, and in the Mount Royal to Mayfair sector. However, the concentration of high negative scores are found in Spruce Cliff, Renfrew, Bridgeland, and Ramsay. The latter areas identify the classic inner city reception areas for immigrants,[22] although the additional concentration associated with the university must also be noticed. Contrasting with this pattern are the high positive scores identifying areas with a high Canadian content, a peripheral suburban distribution in which the older, lower economic status communities of Saddle Ridge and Bowness have the highest values.

The final axis of differentiation recognized in the study accounted for 7.3 per cent of the overall variation, but is much more complex than the other vectors and contains only medium sized loadings. This complexity is reflected in its description as Family Size (East European) - Old Age. The positive side of the vector is dominated by a variety of variables measuring large households and families, dwellings built in the 1960's and with Ukrainians and other East European groups. The negative side, however, is linked with old age characteristics, of people and dwellings, as well as small households, those with female heads and British ethnicity. Figure 7,7 shows that the inner city-suburban distinction again dominates the pattern of scores. The inner city, with an outlier in East Bowness and Saddle Ridge, displays old age characteristics, with the highest values in the communities where senior citizen nursing homes are located, whilst the outer suburbs show the conditions of larger family size.

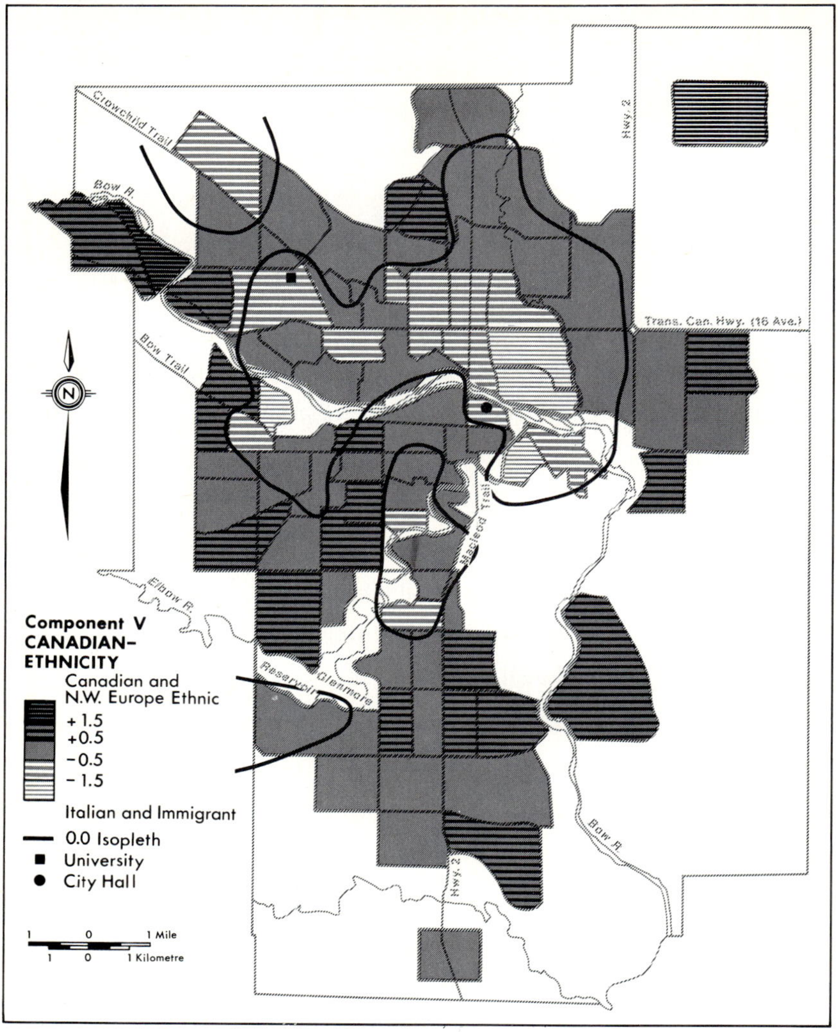

FIGURE 6,7 Distribution of scores on the Canadian-Ethnicity Component.

INTERRELATIONSHIPS BETWEEN AXES AND SCORES

Comparatively few factorial ecologists have gone beyond the description of the factor loading and factor score matrices. However, by re-applying factor techniques to the matrix of correlations between the factors (only possible where oblique axes have been produced) and cluster analysis to the factor scores, even more parsimonious descriptions of the original data set can be produced.

Table 2,7 shows the results of a four axis higher order solution, using the Principal Axes technique and Direct Oblimin rotation with 0.0 as the gamma value. The four axis solution was chosen because a series of tests showed that this was the most concise solution that accounted for over half of the variance of the first order axes. Table 2,7 shows that five of the six axes are linked exclusively with a higher order vector. The clearest and largest association is provided by Migrant Status which forms a separate higher order vector. Components Four (Participation - Non Affluent Status) and Five (Canadian - Ethnicity) load positively on Higher Order Axis I, which means that the impoverishment side of the vector is more closely linked to immigrant status, and high participation is associated with Canadians and those of North Western European ethnic stock. This confirms the traditional views of the process of assimilation. Similarly, the Family Status (Component I) and the complex Family Size (East European) - Old Age (Component VI) axes negatively load on High Order Axis II. This illustrates that the negative sides of Component I and VI, namely non family characteristics and age, are associated with one another. The final high order axis is dominated by the socio-economic status axis, although the Family Size-Old Age dimension is also linked, albeit with a small loading, to the vector. These four higher order axes can be labelled: Family; Social Status; Ethnicity-Affluence; Migrant Status. They confirm McElrath's[23] four axis model of the bases of differentiation of urban areas rather than the more widely used Shevky-

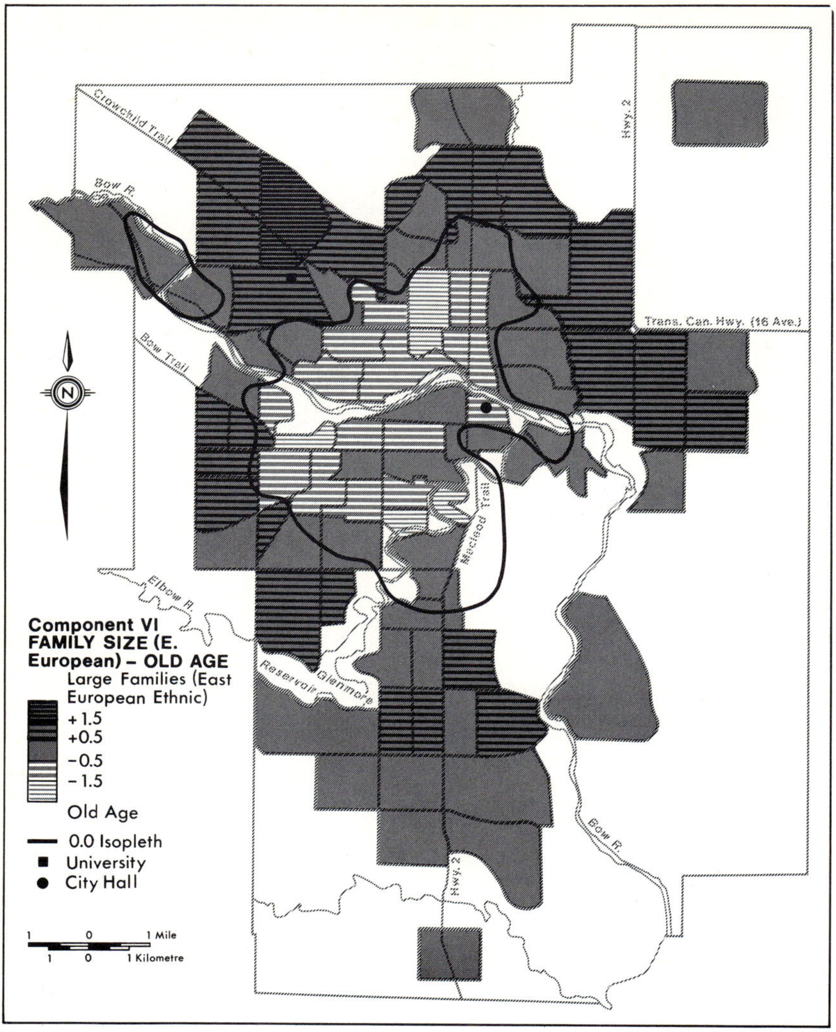

FIGURE 7,7 Distribution of scores on the family size (east European) - old age component.

TABLE 2,7

LOADINGS ON HIGHER ORDER SOLUTION : CALGARY 1971

First Order Axes		Higher Order Axes: Loadings			
				0.3	
No.	Title	III	III	I	IV
1.	Family Status	-88*			
6.	Family Size (E.E) - Old Age	-63	45		
2.	Social-Economic Status		93		
4.	Participation - Non Affluent			73	
5.	Ethnicity			100	
3.	Migrant Status				99

* Decimal points are omitted : -88 is actually -0.88

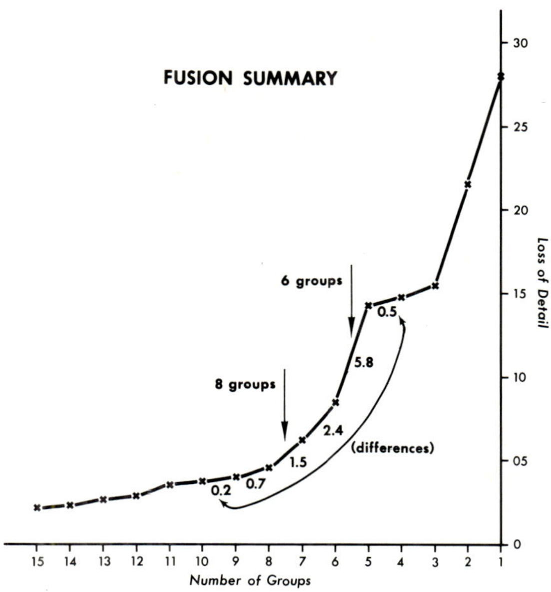

FIGURE 8,7 Fusion summary for Calgary communities.

Bell constructs.

The factor technique used in this essay partitioned the variation in the original data set into separate sources of variation called dimensions, and maps of the scores on these axes showed the individual distributional patterns of these characteristics. By applying cluster analysis to this 83 area x 6 axes matrix the variability of the communities on all six axes can be obtained. Figures 8,7 and 9,7 show the results of a hierarchical cluster routine called Ward's Method[24] to a similarity matrix of Euclidean Distance calculated between the six component scores of each community. Each axis was considered to be a separate source of variation, so they were not weighted according to the size of the axis. As with all cluster methods there is no single solution; the communities are grouped one at a time in a sequential manner until they all form one cluster and at each grouping stage a minimum variance criterion is satisfied. Most cluster users, however, search for discontinuities in the distribution of values measuring the loss of variance at each grouping stage, for example breaks of slope, or large gaps. In Figure 9,7 it can be seen that the first important break of slope occurs after the eighth group, and the biggest gap after the sixth group. These results are used to identify six basic groups, two of them containing a pair of subgroups, to make eight categories in all in the map of Calgary communities (Figure 10,7).

Groups 1,2 and 3 contain most of the communities that made up the built up area of Calgary twenty years ago. The East End community area which is the only representative of Group 1, is characterized by very low Socio-Economic Status, medium Family Status, together with the characteristics of Impoverishment, Old Age and Immigrants. Group 2b, Rosscarrock, Renfrew, Bridgeland, Inglewood, Ramsay, and Victoria, are similar in character, but without such an extremely low level of socio-economic status. With one exception, all the areas are located on the east of the commercial center of the city, the traditional 'zone of

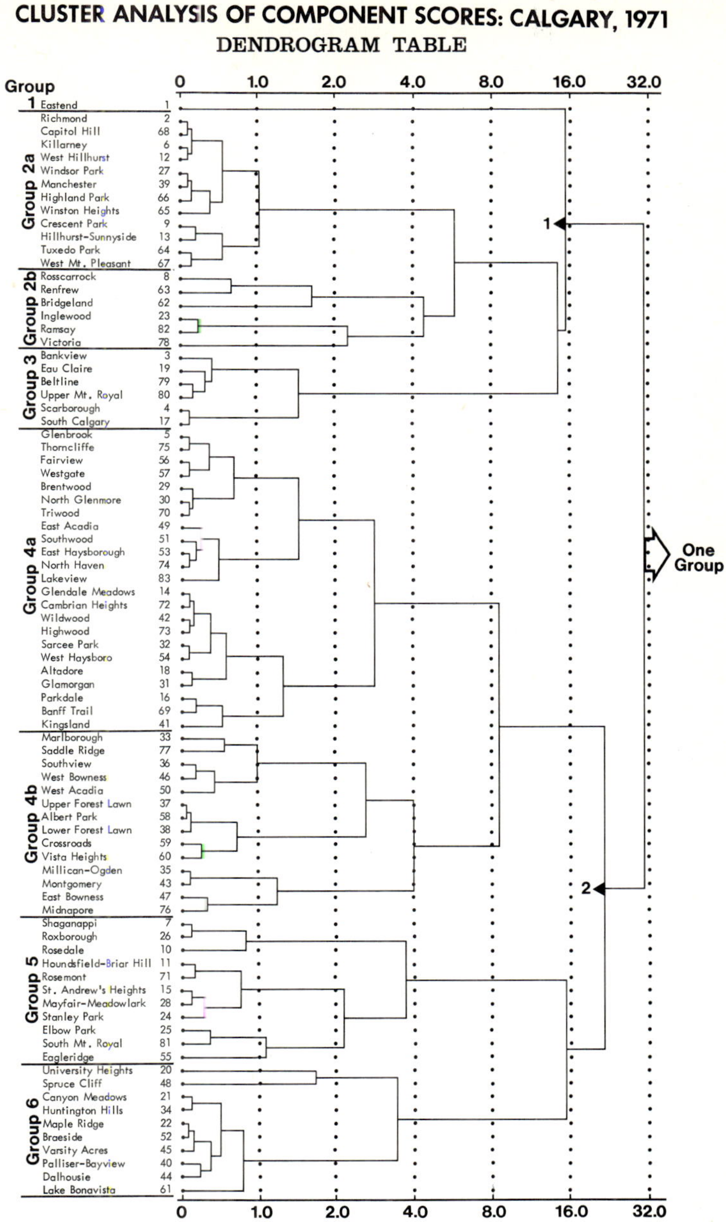

FIGURE 9,7 Cluster linkages for the Calgary communities, 1971.

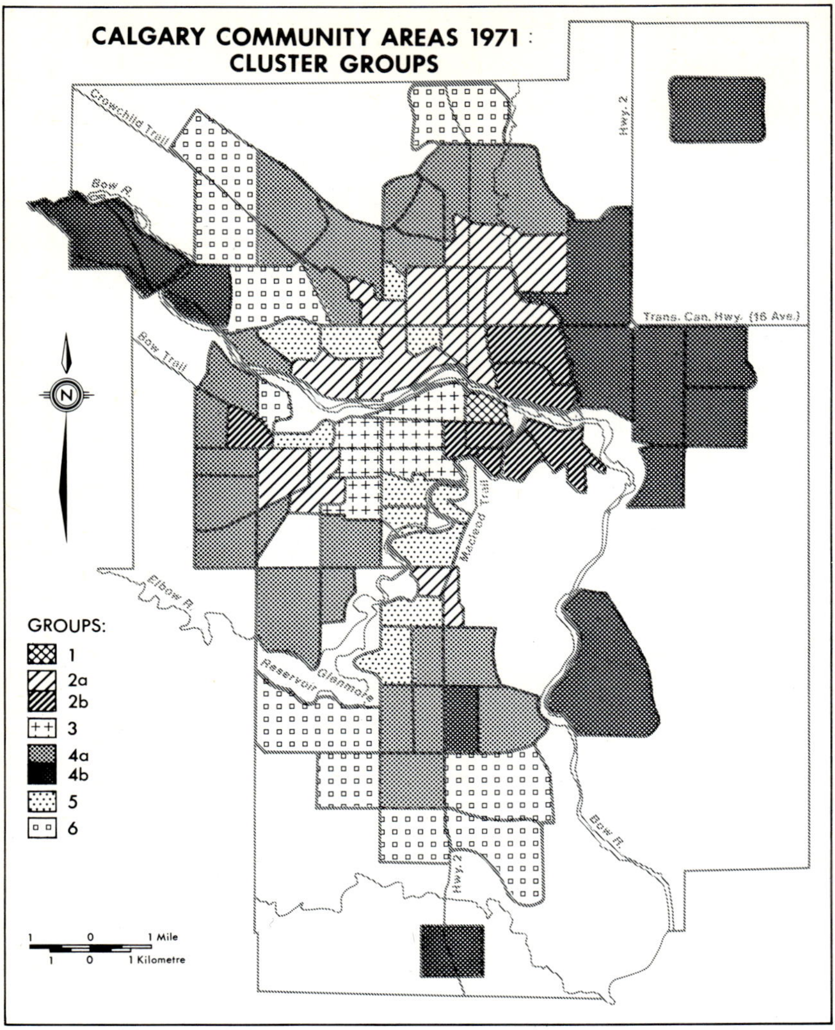

FIGURE 10,7 The distribution of cluster groups of Calgary communities.

discard'[25]. The lack of spontaneous urban renewal led first of all to City Hall intervention via urban renewal projects, and more recently to advocacy planning and community group involvement in the Inglewood-Ramsay area. The exception in this cluster is Rosscarrock, where a large public housing project and the rehousing of senior citizens has produced an area characterized by low socio-economic status, low family status and non-affluence.

Similarly located in the inner city, but this time south of the Bow, in a crescent running from the north to the south of the central area, are the communities comprising Group 3 : Eau Claire, Beltline, Bankview, Scarborough, North Mount Royal and South Calgary. Here some characteristics of older families and poverty can still be found, but this time the community 'mix' is dominated by conditions of very low family status and medium to high socio-economic status. These conditions are associated with the medium and high rise apartment complexes that have grown up on the expanding 'zone of assimilation' on the western edge of the urban core.

Some of the communities of Group 2a have a similar distributional pattern around the inner core, namely West Hillhurst, Sunnyside and Crescent Park, but this time they are located north of the Bow, where the development pressures diffused from the city centre have been less intense. In addition, Group 2a contains a series of other communities at greater distances from the city center, namely Capitol Hill, West Mount Pleasant, Tuxedo Park and Winston Heights on the north, Richmond to the southwest, and Manchester and Windsor Park on the south east. Here redevelopment pressures emanating from the inner city have been less intense, and there are fewer apartment areas, although the barrier once formed by the Bow with respect to the Hillhurst area is quickly being breached. The result is a set of medium scores for the family, migration, and socio-economic status axes and with the comparatively high old age and participation values typical of stable inner city communities but without the extremes of poverty or apartment life found in the zones of discard and assimilation.

Completing the inner city set of communities are those in Group 5. They are distinguished from the others primarily by very high levels of socio-economic status, but contain similar levels of medium to low family status. In accordance with the typical Hoyt[26] model, the distribution forms a south west trending sector running from Mount Royal to Eagle Ridge, although isolated outliers of quite high status areas can be picked out in Shaganappi to the west, and in the St. Andrews Heights - Briar Hill - Rosedale areas to the north of the city.

Of the remaining groups, the most distinct is Group 6. This is composed of the newest communities on or near the suburban periphery that are also of high socio-economic status, such as University Heights, Maple Ridge, Varsity Acres, Dalhousie, Lake Bonavista, Palliser, Braeside. These communities are characterised by high family status, quite large families, and high migration levels. A combination of all these characteristics means that the group also contains middle income or middle to high income communities such as Huntingdon Hills, Spruce Cliff, and Canyon Meadows. Groups 4a and 4b contain a more complex set of suburban places, with the characteristics of high family status, high levels of migration, large families and high participation levels. The division between them is primarily one of socio-economic status. In Group 4b are the lower income communities of the eastern side of the city, from Vista Heights and Forest Lawn to Millican, Ogden and West Acadia and with outliers in the Bowness-Montgomery area and at Saddle Ridge and Midnapore. Group 4a is comprised of the remainder, and these are concentrated in three areas : the northern suburbs from Thorncliffe through Cambrian Heights and Banff Trail to Brentwood; in the south west from Westgate to Glamorgan, North Glenmore and Altadore; and in the south around Macleod Trail, from Haysborough and Kingsland on the West, to Fairview and Acadia on the east.

CONCLUSIONS

The geographica description of the social character of any urban area should deal wi-h two distinct features : firstly the identification of the basic clusters of variation found within the urban complex, and secondly the way in which these clusters are distributed within the urban area. This study of Calgary has shown that there are four major sources of variation in the data set used : Family Status; Economic Status; Migrant Status and Ethnicity. Two additional minor axes of differentiation were also produced, a rather complex Family Size - Old Age vector and a Participation - Impoverishment factor. The latter seems to be the most important judging by comparisons made with other studies and from the results of the higher order analysis. This means that the social dimensions of the community areas of Calgary conform very closely to the models proposed for other western cities, despite the local and individual peculiarities of Calgary's growth.

In distributional terms, each community can be scaled along the six axes, and again most of the standard findings are confirmed. Family status shows a concentric pattern whilst the Socio-Economic axis splits the city into an east and west zone, with a sector of low status extending westwards along the Bow Valley to the older industrial communities of Bowness and Montgomery, areas that were only annexed by the city a decade ago. In addition, Migration status is linked to the growing edge of the city, and to the areas of transformation around the commercial centre. The Ethnicity axis shows a more clustered pattern, but the results show that Calgary does not have any large concentrations of ethnic groups comparable to the French in Winnipeg or the Ukrainians in Edmonton, at least not at the community level analysed in this study. Finally, the Family Size-Old Age axis displays an additional inner city-suburban differentiation, whilst the late family or dispersing family groups in the inner suburbs show high levels of work-force participation and young adult characteristics. This implies a crystallization of the axis out of the family status vector, a process different from that which left behind the impoverished groups when the majority gained in economic status. In general, however, the higher order links of the bi-polar Participation - Impoverishment vector with the Ethnicity vector demonstrates the way in which ethnicity is only one of the ways in which groups are separated out from the mainstream of society.

The results of this study show that the social dimensions and patterns of the community areas of Calgary are very similar to those of other Western cities. They provide a more satisfactory basis for future studies of the particular idiosyncracies of the city by providing a descriptive basis of the norm, from which the degree of deviation can be calculated. But this description of the particular results of the analysis for Calgary is not the only objective of this chapter. It is necessary to return to the methodological points discussed at the beginning of the chapter and ask whether these results could not have been achieved more simply. As with all factor and cluster analysis studies, two problems emerge. The first is that the results may seem 'obvious' to a Calgary planner, developer or citizen. The second is that they provide little evidence of the flavour and variety of urban life in the urban communities.

With respect to the first point, the fact that the results correspond to commonsense experience is an advantage rather than a disadvantage. It demonstrates that the description of communities need not rely upon a lengthy experience of living in Calgary, but that it can be derived from the application of standard synthesizing techniques to a census data set. Hence we have described a method that produces results that could be obtained by an individual located in Ottawa or London, one without experience of living in Calgary, although a knowledge of Calgary clearly represents an important check on the results. Of greater importance is the fact that although the final grouping of communities may be a familiar one, it is much more unlikely that such a succinct description of the axes of differentiation can be produced. If it can, it is the product of a rare ability. So if we can replace this intuition by statistical manipulation, then we may have released that intuitive effort for other tasks. Finally, it must be noted that the method by which the results were obtained, ccn be easily extended to other cities. This produces a rigorous quantitative comparison that goes far beyond any subjective characterization of places.

The answer to the second point is much less satisfactory. The results derived in this type of study are completely dependent upon the utility of the initial data set, and normally require the compilation of 'hard' statistical data. Unfortunately, the process of obtaining these data moves us away from what might be called the 'essence' of place, from the reality of sights, smells, and the variety of life in the various communities. Although recent developments in urban perception bring back some of the spice of urban life, it is difficult to see how all of them can be accommodated in a single study. Some of the features may be included, but to encompass the totality of community characteristics we may have to resort to investigative reporting as an adjunct to the type of approach used in this study. Yet at the very least this type of description has the merit of dealing with a defined data set. Hence we know how much of the reality of place is being dealt with, in this case social structural conditions. Since the study is capable of replication, unlike the older descriptive approaches, a systematic body of generalizations on the dimensions and patterns of urban places can be produced. To use an analogy we can conclude by observing that the problem of describing cities is rather like dealing with Humpty Dumpty. We all have some idea of what Humpty Dumpty looks like, but once he falls and is split into pieces, how do we put the pieces back together again in a systematic manner? And even if this is achieved, then can we relate Humpty Dumpty to other models? It is a task of urban geographers who seek to develop systematic and testable generalizations to attempt this task, even if the first results are little better than those obtained by simpler means.[27]

APPENDIX I

LIST OF VARIABLES USED IN THE STUDY

(Short titles are followed by a detailed description)

AGE

1. Children. Percentage of the Total Population that are under 14 years of age.

2. Young Adults. Percentage of the Adult Population between 15 and 24 years of age.

3. Mature Adults. Percentage of the Adult Population between 25 and 44 years of age.

4. Middle Age. Percentage of the Adult Population between 45 and 65 years of age.

5. Old Age. Percentage of the Total Population over 64 years of age.

FAMILY

6. Single Adults. Percentage of the Adult Population that are single.

7. Divorced. Percentage of the Adult Population who have been divorced.

8. Family Size. Average number of persons per family.

9. Number of Children. Average number of children per family.

10. Lodgers. Percentage of the households with lodgers.

FEMALE CHARACTERISTICS

11. Females. Ratio of the total number of females to males.

12. Fertility ratio. Ratio of the number of children to fertile females (aged 15 - 44 years old).

13. Female heads. Percentage of the total number of households having female heads.

14. Female participation. Percentage of adult females who are in the labour force.

TYPE AND TENURE OF DWELLINGS

15. Owner occupied. Percentage of the dwellings that are owner occupied.

16. Single detached. Percentage of the dwellings that are single detached.

17. Apartments. Percentage of apartments in the total number of dwellings.

18. Substandard. Percentage of dwellings without hot running water, without exclusive bath or shower, or without exclusive use of a toilet.

HOUSEHOLDS AND DWELLING SIZE

19. Small dwellings. Percentage of the dwellings with four or less rooms.

20. Large Dwellings. Percentage of the dwellings with eight or more rooms.

21. Persons per household. Average number of persons per household.

22. Small households. Percentage of households with only 1 or 2 persons.

23. Large households. Percentage of households with more than 7 persons.

PARTICIPATION AND EDUCATION

24. Participation ratio. Percentage of the Adult Population who are in the labour force.

25. Young adult participation. Young adults in the labour force as a percentage of the young adult population.

26. Unemployed. Unemployed as a percentage of the employed population.

27. Limited education. Percentage of adults who have less than grade 9 education.

28. High School. Percentage of adults with Grade 11-13 education .

29. Graduates. Percentage of adults who have degrees.

OCCUPATION AND INCOME

30. White Collar Ratio. Ratio of White Collar occupations to all other occupations.

31. Managerial. Percentage of the labour force in managerial occupations.

32. Vocational. Percentage of the labour force in teaching, medicine or technical occupations.

33. Manufacturing. Percentage of the labour force in processing or machinery occupations.

34. Construction and Transport. Percentage of the labour force in construction and transport occupations.

35. Household Income. Average household income per annum.

MIGRATION

36. Non Migrants. Percentage of the population over 5 years of age who have not moved houses.

37. Low Intermunicipal Movers. Percentage of people who have only moved once between cities.

38. High Intermunicipal Movers. Percentage of movers between cities who have moved more than three times.

39. Immigrants. Percentage of the total population born outside Canada.

40. Recent Immigrants. Percentage of the Immigrants who have moved since 1966.

41. Provincial Immigrants. Ratio of the population born in a different province from their present location to those born in the province.

ETHNICITY

42. Canadian Stock. Percentage of the population with both parents born in Canada.

43. Canadian citizens. Percentage of the population who are Canadian citizens.

Percentage of the population from the following ethnic stocks:

44. British (British Isles including Ireland)

45. French

46. German (Germany and Austria included)

47. Ukrainian

48. Asiatic

49. Native (Indian and Eskimo)

50. Jewish

51. Netherlands

52. Scandinavian

53. Italian

54. Hungarian

55. Polish

AGE OF CONSTRUCTION

56. Pre 1946. Percentage of the total number of dwellings built prior to 1946.

57. 1946–50. Percentage of the total number of dwellings built

between 1946 and 1950.

58. 1951–60. Percentage of the total number of dwellings built between 1951 and 1960.

59. 1961–65. Percentage of the total number of dwellings built between 1961 and 1965.

60. 1966–70. Percentage of the total number of dwellings built between 1966 and 1970.

REFERENCES

1. WIRTH, L. "Urbanism as a Way of Life," American Journal of Sociology, 44 (July, 1938), pp. 1-24.

2. MAYHEW, H. London's Labour and the London Poor. Vols. 1-3, Vol. 4, 1862. London : Griffin Bohn and Co.

3. ENGLES, F. The Condition of the Working Class in England. Chicago: Stanford University Press, 1958.

4. WOODSWORTH, J.S. My Neighbour : A Study of City Conditions, a Plea for Social Services. Toronto: (Reprinted by) University of Toronto Press, 1972.

5. BOOTH, C. Life and Labour of the People of London, 17 Volumes. London : Macmillan, 1902.

6. BOOTH, C. "Life and Labour of the People of London: Inquiry based on the 1891 Census", Journal of the Royal Statistical Society, No. 56 (Dec. 1893), pp. 557-596.

7. SHEVKY, E. and BELL, W. Social Area Analysis, Stanford University Press, 1955.

8. PARK, R.E. et al. (eds.) The City. Chicago: University of Chicago Press, 1925 (reprinted 1967). See in particular: Burgess, E.W. "The Growth of the City", pp. 47-62.

9. HARMAN, H. Modern Factor Analysis. Chicago: University of Chicago Press, 1968.

10. SNEATH, P. and SOKAL, P. Numerical Taxonomy : The Principles and Practice of Numerical Classification. San Francisco: W.H. Freeman, 1973.

11. RUMMEL, R.J. Applied Factor Analysis. Evanston : North Western University Press, 1970.

12. DENT, O. and SAKODA, J.M. "Potential Sources of Spuriousness in Factor Ecology Studies", Institute of British Geographers Quantitative Methods Study Group Working Paper No. 1 (1973), pp. 37-44.

13. DAVIES, W.K.D. and BARROW, G. "A Factorial Ecology of

Three Prairie Cities", The Canadian Geographer, 17, No. 4, (1973), pp. 327-353.

14. DIXON, W.J. Biomedical Computer Series. University of Los Angeles, 1968. Checks of the accuracy of these package programmes were made by also using: Nie, N., Brent, D. H. and Hull, C.H. Statistical Package for the Social Sciences. New York: McGraw Hill, 1971.

15. SHEVKY, E. and BELL, W. op. cit.

16. MURDIE, R.A. The Factor Ecology of Metropolitan Toronto, 1951-1961. Department of Geography, University of Chicago, Research Paper No. 116, (1969).

17. TIMMS, D. Urban Mosaic : Towards a Theory of Residential Differentiation. University of Cambridge Press, 1971.

18. ENGELS, F. op. cit.

19. BURGESS, E.W. in PARK, R.E., op. cit.

20. MURPHY, R.A. and VANCE, J.E. "Internal Structure of the Central Business District", Economic Geography 31 (1955), pp. 21-45.

21. DAVIES, W.K.D. and LEWIS, G.J. "The Urban Dimensions of Leicester", in Gleave, B. and Clark, B. (eds.) Social Patterns of Cities. Institute of British Geographers Special Publication No. 5, (July 1973), pp. 71-86.

22. BURGESS, E.W. in PARK, R.E. op. cit.

23. McELRATH, D.L. "Societal Scale and Social Differentiation: Accra, Ghana", in Greer, S. et al. (eds.), The New Urbanization. New York: St. Martin's Press, 1968.

24. WISHART, D. "Fortran II Programs for Eight Methods of Cluster Analysis". Kansas Geological Survey Computer Contributions, No. 38 (1968).

25. MURPHY, R.A. and VANCE, J.E. op. cit.

26. HOYT, H. The Structure and Growth of Residential Neighborhoods in American Cities. U.S. Federal Housing Administration, Washington, D.C., 1939.

27. This study is part of a large project on the comparative factorial ecology of Prairie cities. The financial support provided to the project by Canada Council Grant Number S73-1185 is gratefully acknowledged.

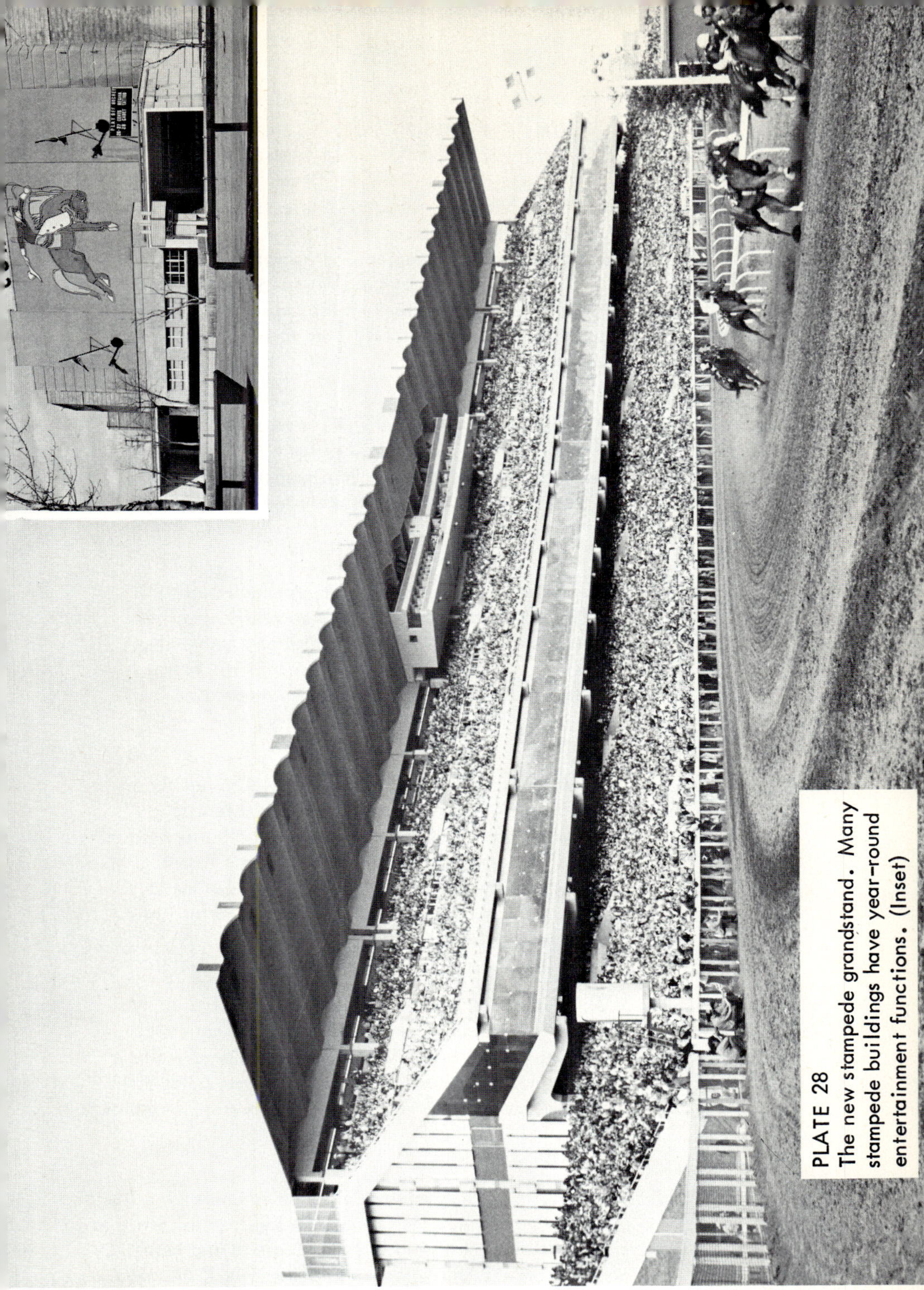

PLATE 28
The new stampede grandstand. Many stampede buildings have year-round entertainment functions. (Inset)

Calgary Exhibition and Stampede Photo

THE CONTRIBUTORS

Brenton M. Barr, B.A., M.A., Ph.D., Associate Professor, Department of Geography, University of Calgary. Before joining the Department of Geography at Calgary, Dr. Barr taught for three years at the University of Alberta. His first two degrees are from the University of British Columbia, and his doctorate from the University of Toronto.

Wayne K.D. Davies, B.Sc., Ph.D., Professor, Department of Geography, University of Calgary. Dr. Davies chief research interests are the methodology of Geography and urban geography. Both his degrees were awarded by the University of Wales.

Timothy T. Gyuse, B.A., M.A., Assistant Lecturer, Ahmadu Bello University, Nigeria. Mr. Gyuse is currently enrolled in the Faculty of Environmental Design, University of Calgary while on leave of absence. His first degree is from the Ahmadu Bello University and his second from the University of Calgary.

Denis Johnson, B.A., M.A., Ph.D., Associate Professor, University of Alberta. Dr. Johnson's chief research interests are urban geography and the application of film technique to geographic research and teaching. His first degree was awarded by the University of British Columbia, his second by the University of Alberta and his Ph.D. by Queen's University, Belfast.

Alice M. McEwen, B.A., Graduate Student, University of Calgary. Ms. McEwen's research interests are regional economics, industrial location and manufacturing linkage. She is currently working towards her M.A. degree, her B.A. being awarded in 1971 by the University of Minnesota, Duluth.

Peter J. Smith, B.A., M.A., Ph.D., Professor and former Chairman, Department of Geography, University of Alberta. Dr. Smith received his early training in New Zealand and his Ph.D. degree from the University of Edinburgh in 1964. His current research is into the impact of local government decision-making, particularly planning decisions, on the urban environment.

Don G. Harasym, B.A., B.Ed., M.A., Research Officer, Conservation Authority, Alberta. Mr. Harasym received his B.A. from the University of Calgary in 1970, his B.Ed., from the University of Alberta in 1972 and his M.A. from the same institution in 1975.

George H. Zieber, B.A., B.Ed., M.A., Ph.D., Associate Professor, University of Lethbridge. Dr. Zieber obtained his B.A., B.Ed., and Ph.D. degrees from the University of Alberta and his M.A. from the University of Toronto. Prior to his appointment at Lethbridge he was a geographer-analyst with the Ontario Department of Highways (1962-1964). During the years 1964-1968 and 1972-1973 he was the Acting Chairman, Department of Geography, Lethbridge Junior College and University of Lethbridge.